UNIPA Springer Series

More information about this series at http://www.springer.com/series/13175

Maria Laura Scaduto

River Contracts and Integrated Water Management in Europe

UNIVERSITÀ
DEGLI STUDI
DI PALERMO

Springer

Maria Laura Scaduto
University of Palermo
Palermo
Italy

ISSN 2366-7516 ISSN 2366-7524 (electronic)
UNIPA Springer Series
ISBN 978-3-319-42627-3 ISBN 978-3-319-42628-0 (eBook)
DOI 10.1007/978-3-319-42628-0

Library of Congress Control Number: 2016946930

Printed on acid-free paper

This Springer imprint is published by Springer Nature
The registered company is Springer International Publishing AG Switzerland

The care of rivers is not a question of rivers,
but of the human heart.

Tanako Shozo

Foreword 1

In recent years, the senseless human interventions and climate change on a global scale have contributed to the intensification of extreme weather events and exceptional natural phenomena that, in addition to highlight the fragility of the territory and particularly of ecosystems closely linked to river basins, represent serious threats to the safety of populations. Indeed, although there are many different planning tools including those concerning river basins, we are faced with a territory not yet fully planned and still too exposed to the impact of historical anthropic processes, such as illegal building, water pollution and landscape alteration.

The large number of plans and programs of diverse nature, managed by different subjects, their low level of integration and the scarce degree of the community participation, very often returns images and realities of territories not yet adequately planned and, therefore, not prepared to cope with extreme climate changes, as well as natural and socioeconomic evolutionary processes.

The attitude of different countries dealing with such global issues was different in time and in terms of adopted instruments. For example, it is well known the advantage position of France that, since the early 1960s, has recognized the importance of planning at the river basin scale, identified as the optimal territorial unit for the integrated management policies. Therefore, policies and regulations specifically addressed to plan and safeguard territories have been put in place in the 1980s. The dissemination of *contrats de rivière* inserted in this evolutionary scenario as a result of a long season of negotiated and participated practices of water resources and river territory management. In Italy, the same issues have been dealt with similar instruments only in the last decade, through the activation of first river contracts and the recognition on part of many institutions of the importance to adopt them as new tools for both water resources management at the river basin scale and potential integration of different spatial planning levels.

In light of these premises, the river contract appears the most suitable instrument for such purposes as it promotes voluntary agreements between public institutions and private individuals, new forms of institutional cooperation, consultation and participation, as well as new ways of integrating the different practices of spatial and sectoral planning. In particular, within the Italian scenario, characterized by its low coordination degree between different planning competences and tools, river contracts have taken an intermediate position between river basin and water management plans, on the one hand, and regional and local spatial plans, on the other.

With regard to such wide and complex themes, the research illustrated in this volume by Maria Laura Scaduto offers an updated overview of the European legislative and procedural scenario, a comparative analysis of the two paradigmatic cases of France and Italy, and an examination of the main application experiences of river contracts and their outcomes. For its well-structured theoretical, methodological and procedural contents, this volume is aimed at a wide and varied public relating to research community, public and private institutions, professional sector and citizenry, in line, therefore, also with the principles of *participation* and *knowledge sharing* expressed by the *Integrated Water Resource Management* paradigm.

The research work clearly shows the complexity of ecosystems linked to river basins, within which ecological instances and different uses of water resources are still to be better harmonized, conflictual situations are continuously emerging, while new opportunities for shared projects between public and private actors are arising. In response to these issues, river contracts have emerged as dynamic and versatile tools that can help overcome the misalignment between different planning levels, achieve the balance of socioeconomic development and natural resources safeguard, in particular of water resources, and promote new synergies between public and private actors, and the community participation in the design and planning decisions.

In this perspective, the comparative analysis undertaken between France and Italy, taking into proper account their differences in terms of territorial and administrative characteristics, offers two complementary levels of thematic reading about integrated water management policies and river contract adoption. The comparison is underpinned by the examination of four river contract case studies activated within significant river basins, two of which located in metropolitan areas (*Contrat de Rivière de l'Yzeron*, in France; *River Contract of Olona-Bozzente-Lura*, in Italy) and two other initialed within river basin predominantly characterized by rural territories (*Contrat de bassin de la Basse Vallée de l'Ain,* in France; *Ofanto Valley River Contract*, in Italy).

On the whole, this volume explicitly illustrates to which extent river contracts emerged as innovative programming and planning tools, often overcoming institutional and legal competence conflicts, and are revealing as dynamic paths capable to activate the desirable integration process between river basin and spatial planning, and to support new forms of public participation in territorial governance.

Prof. Ignazia Pinzello
Full Professor of Urban and Regional Planning
University of Palermo
Palermo, Italy

Foreword 2

This research work by Maria Laura Scaduto puts into perspective the over thirty years of European policies aimed at improving water management practices. Particularly, it illustrates every effort made to achieve actual integration at the river basin scale among the, as yet, overly sectoral management approaches.

However, some will object that many European practices have been conducted in an integrated manner for quite a while, at the hydrographic basin scale as well as at the local management level. For illustrative purposes let us consider two examples, so as to better illustrate their limits.

In France, a number of mountain slopes (Alps, Pyrenees or Apennines) underwent intense erosion phenomena in the eighteenth and nineteenth centuries, due to excessive deforestation and overexploitation of both pasture and grain crops, thereby weakening the soil in a difficult climatic context, namely that of the Little Ice Age. The widespread flooding and damages in the valleys raised awareness of the mismanagement of mountains and the need for upstream–downstream integration of practices. Starting from the 1830s, reforestation policies as well as a corollary eviction of the rural population, considered excessive, were promoted. This policy was actually put into place starting in the 1860s on the basis of binding legislative frameworks. Indeed, these policies were conducted at the hydrographic basin scale and to better manage rivers and streams, but, as those practices were designed and implemented in an authoritarian manner, they failed to take into account the needs and wishes of the concerned communities. The slopes were *treated*, and erosion was reduced, but the mountains were emptied of their populations.

The integration of management methods was also attempted and achieved at the scale of valley section. Let us take another French example, that one of the Gave de Pau, at foot to the Pyrenees. In the late 1960s, the policy pursued by state services at the local level was focused to gravel mining of riverbeds. Why? Because the entrepreneurs of quarrying sector would have favored the construction of granulated-based structures and embankments for public works, flooding would be mitigated and farmers would have enjoyed improved conditions for production. It would have sufficed to erect dikes to keep lateral erosion under control and weirs along the river to control the vertical erosion. Although the goal seemed beneficial for the economy and some actors in the territory, the outcome was severely (albeit vainly) criticized by the Ministry of Environment in the 1980s, and because flooding was exacerbated downstream of the 20-km river segment concerned by interventions, the alluvial forest languished and alluvial groundwater had lost a considerable part of its capacity. It lacked the upstream–downstream (or basin) perspective and the higher-order features of what we now consider truly integrated management. One could bring countless European examples of such interventions on river banks, systematically undertaken to protect one particular interest or another. Even though a river contract for the Gave de Pau in the Pyrenees was in effect as far back as 2002 (upstream, in the zone of Lourdes), none exists for the Pau region, nor is there any *Schéma d'Aménagement et de Gestion des Eaux* at the river basin scale. Evidently, the bottom line is that nowadays a lot of work still remains to be done.

In some ways, the situation changed in Europe and peculiarly in France in the mid-1970s. In those changes, one should recognize the often implicit conjunction of circumstances, such as raised awareness, research works and perhaps of the general scenario, as disjointed yet synergistic elements that favored a paradigm shift. Maria Laura Scaduto reminds us that the outcomes of the *Mar del Plata Conference* of 1977, which favored integrated water resource management, arguably ahead of its time, carried over to the *Dublin Conference* (1992) which finally formalized the essential principles commonly accepted nowadays.

What happened in Europe and elsewhere in the world during these fifteen long years? Let us remain within this context characterized by some factors that by no means encompass the issue in its entirety. In 1978, the research, although not limiting the discussion to this, concerning the analysis of aquatic ecosystems was officially launched specifically to understand how to harness impact studies, so as to come up with actual supporting tools to manage burdensome interventions in the water domain. The contribution of fluvial geomorphology became a necessity, benefiting from the works undertaken on the *fluvial system* defined in particular by American geomorphologists. There is quite compelling evidence that research is too complex to be addressed without interdisciplinary efforts, if what we pursue is the effective integration of disciplines. Despite the many attempts, the opening to the humanities still remains limited, whereas the *Agences de l'Eau*, government bodies and services, as well as some managing organisms, are very keen on paradigm shifts. And they are not alone because the social body is being profoundly changed in a period of highly controversial, non-environmentally friendly, management approaches. NGOs will play a very important role as intermediaries between science and public opinion in a sociopolitical system that decompartmentalizes itself and promotes so-called *citoyennes*, i.e., decentralized participatory and community-driven practices.

To what extent do river contracts, introduced in France in the early 1980s, reveal themselves as innovations in policy that break with previous practices? Firstly, as this volume duly highlights, by replacing the, too frequently, partial and sector-driven state policies, in countries characterized by strong centralization, with the practice of stipulating contracts between partners at the basin scale. These new contractual agreements strive to reconcile economic development, based on multiple uses of water, with values that are emerging and being recognized, such as the social uses and ecological quality of the environment. Therefore, it is necessary to define development on a basis that integrates multiple interests at the scale of territorial systems, no longer based on traditional administrative boundaries, but rather on spatial entities based on water territories. Once chosen the river basin scale, there remains to pursue the implementation of new sectoral practices which must respect the coherence of multiple interests. The decentralization of institutional competences becomes a major issue that will determine the success of the project. In France, on the basis of the experience acquired over nearly a decade, the *Assises de l'Eau* (1990) constituted a forum that was a prelude to the *Loi sur l'Eau* of 1992. The *Schémas directeurs de bassin*, from which progressively stemmed the *Schémas de Gestion et d'Aménagement des Eaux* (SAGE), afforded an even wider (and much needed) coherence to those initiatives represented by *contrats de rivière*. In this perspective, also the *European Water Framework Directive* (2000) is harmonizing existing and future practices and represents an effective tool to incentivize these forms of integrated water management.

What brought these innovative approaches to the specialists of river basin functioning, through the interdisciplinary perspectives of hydrology, geomorphology and ecology?

First of all, the river contract provided the possibility *to implement concrete management practices, built on integrating concepts*. The period from the early 1980s to the early 1990s was that of the passage from the *scientific integrator concept* to the forms of integrated management, which are hardly the same thing. The valley of the Rhone River was thus the setting of the preparation and experimentation of the scientific concept of the *fluvial hydrosystem*. It was subsequently the site of its implementation through collaborations between the *Compagnie Nationale du Rhône* and the *Agence de l'Eau Rhône-Méditerranée et Corse*, then also involving local authorities and communities. This was achieved on the Rhône and its tributaries within a framework consisting of *contrats de rivière*, the elaboration of a SAGE and the ensuing *plans Rhône*.

These principles provide a scientific basis for the approach based on the analysis of environmental conditions, which must be clearly expressed and understood by all stakeholders. Let us consider, for example, streams in basins comprising mountainous regions or even hills. Nowadays, a frequent management issue to be addressed is the sinking, i.e., vertical erosion, of rivers due to *sediment deficit*. The key concepts are those of *sedimentary cascade* and *sediment budget*. The first analyzes how slopes produce sediment by erosion, how it is stored at the bottom of the slopes or reach the riverbed (concept of slope–riverbed coupling) and how it is moved downstream or is retained at natural or artificial sites. The latter concept that is of sediment budget quantifies these factors and accurately locates the points where that action is desirable. It behooves us to define the nature of that action. As can be seen, the concepts provide a cogent and replicable framework, based on concrete realities that inhabitants can observe by themselves, even without quantifying them.

Secondly, the purview of river contracts (usually a hydrographic basin) is amenable to territorial management support through the application of hydro-ecomorphological concepts. *The areas of scientific analysis and management overlap*. Why is this so important? The scope and application of hydraulic engineering works have traditionally been restricted to fluvial sections (eroded banks, weirs and river groynes to offset excessive drive or threats to bridges or dams, etc.). These recesses are too limited because they fail to take into account river continuity. Designing the *hydrographic system* with reference to a river basin tends to ensure that the intervention on the river system subsumes the selection of sites actually relevant with respect to the interventions. This entails passing from a localized approach to management, thereby merely tailoring issues to local applications, to more rational forms of management that avail themselves of the teachings of sediment budget. In other words, the banks exposed to the erosion are not protected in a hard (or soft) way if they are located in a river section that is in *sediment balance* (i.e., where the material outputs and inputs are equivalent). The lateral river erosion is the manifestation of balance and instead of intervening to block the process, it will be best to innovate in favor of new practices, such as the purchase of land to anticipate erosion. The result will be both effective and sustainable.

The methods of implementing river contracts, as shown by the fine work by Maria Laura Scaduto, provide the key to access these new more sensible and *citoyen* water management modes, meaning by this that it is possible and desirable to more directly involve basin populations (not only the inhabitants concerned by the erosion of the main river or stream banks) in the design process and then in the political and management decisions. It is a profound paradigm shift indeed. To some extent, nowadays in Europe we find again the principles of hydro-sedimentary functioning that had inspired old restoration policies in the context of mountain land. The major innovation consists in the nature of the political approach: no longer imposing compulsory measures are dictated by the state government, but rather educating the citizenry, while providing them with operational tools and inviting them to actively participate in the decision-making process.

Prof. Jean-Paul Bravard
Professor Emeritus of Geography
University Lumière Lyon 2, France

Acknowledgments

The Author is very grateful to Prof. Ignazia Pinzello and to Prof. Jean-Paul Bravard for all the support guaranteed (respectively, in her role of tutor and his role of co-tutor) during the three-year work period that led to the final discussion of the Ph.D. thesis *Planning fluvial territories. The river contract as an instrument for an integrated management at the river basin scale*, elaborated within the 23rd Cycle of Research Doctorate in Regional and Urban Planning (2009–2011), at the University of Palermo, with a co-tutoring of the French University *Lumière-Lyon 2.*

Particular acknowledgments are due to Mario Clerici (General Directorate of Environment, Energy, Sustainable Development of the Regional Administration of Lombardia); Mauro Iacoviello (Agency for Territorial Pact of the Province of Bari and Ofanto Valley; Provincial Administration of Barletta-Andria-Trani); Alberto Magnaghi (University of Firenze); Massimo Bastiani (Coordinator of the Italian Board on River Contracts; University of Rome Sapienza); Andrea Scianna (Italian National Research Council; University of Palermo); Stéphane Guerin (Syndicat d'Aménagement et Gestion de l'Yzeron, du Ratier et du Charbonnier); Jean-Philippe Ravasseau and Céline Thicoïpé (Syndicat de la Basse Vallée de l'Ain). During these years, they have offered many precious contributions to the reflections developed firstly within the Ph.D. thesis and later deepened and illustrated in this volume.

Finally, the Author wishes to thank the publisher Springer and the editorial board of UNIPA Springer Series, for having offered me the opportunity to contribute to this prestigious book series, in particular Prof. Carlo Amenta.

Contents

Abbreviations

AATO	*Autorità d'Ambito Territoriale Ottimale*
ARPA	Regional Environmental Protection Agency
ATO	*Ambiti Territoriali Ottimali*
CdF	*Contratto di Fiume*
CdR	*Contrat de Rivière*
CLE	*Commissionn Local de l'Eau*
DREAL	*Direction Régionale de l'Environnement, de l'Aménagement et du Logement*
DTA	*Directive Territoriale d'Aménagement*
EC	European Community
GIS	Geographical Information System
ICT	Information Communication Technologies
IWRM	Integrated Water Resource Management
LEMA	*Loi sur l'Eau et les Milieux Aquatiques*
NTA	*Norme Tecniche di Attuazione*
PAGD	*Plans d'Aménagement et de Gestion Durable*
PAI	*Piano per l'Assetto Idrogeologico*
PCDN	*Plan Communal de Développement de la Nature*
PCDR	*Programme Communal de Développement Rural*
PCEDD	*Plan Communal d'Environnement pour le Développement Durable*
PGD	*Piano di Gestione del Distretto idrografico*
PLIS	*Parchi Locali di Interesse Sovra-comunale*
PLU	*Plan Local d'Urbanisme*
PPGIS	Public Participation Geographical Information System
PPR	Regional Landscape Plan
PPRI	*Plans Prévention des Risques Inondation*
PRTA	Regional Water Protection Plans
PTCP	Territorial Plan for Provincial Coordination
PTPR	Regional Landscape and Territorial Plan
PTR	Regional Territorial Plan

RBM	River Basin Management
RC	River Contract
SAGE	*Schéma d'Aménagement et Gestion des Eaux*
SAGYRC	*Syndicat d'Aménagement et Gestion de l'Yzeron, du Ratier et du Charbonnieres*
SCoT	*Schéma de Cohérence Territoriale*
SDAGE	*Schéma Directeur d'Aménagement et Gestion des Eaux*
SEAGYRC	*Syndicat d'Etude pour l'Aménagement et la Gestion de l'Yzeron, du Ratier et du Charbonnières*
WFD	Water Framework Directive

Chapter 1
Theoretics and Methodology

> The care of rivers is not a question of rivers,
> but of the human heart.
> *Tanako Shozo*

Abstract Since 2000 in Europe an integrated management framework has been developed to innovate exploitation and safeguard of water resources. In this context the EU Water Framework Directive has identified the hydrographic basin as the optimal territorial unit for promoting new participatory policies, based both on the interaction of stakeholders and the coordination of sectorial instruments. In this scenario, river contracts assumed a strategic role both in addressing these purposes and supporting the dialogue and integration between interests of public and private stakeholders. This chapter illustrates the theoretical and methodological framework, and the comparative approach on which the research work has been based to evaluate the effectiveness of river contracts and their relationships with urban and territorial planning.

1.1 Introduction

Since 2000, the European Community (EC) has been developing an integrated water protection framework and promoting the orientations in terms of the exploitation and safeguard of water resources and soil, identifying the hydrographic basin as the optimal territorial unit for their management (EC 2000, 2007).

The underlying priorities are the involvement and participation of stakeholders, and the coordination and integration of current sectorial instruments and policies, also in line with the paradigm of *Integrated Water Resource Management* (IWRM) (GWP 2000).

In this scenario, key processes are the analysis, monitoring and updating of regulatory and practical instruments. Among the latter, particularly at the scale of the hydrographic basin, the *river contract* (RC) assumed a strategic significance for its great potential in the integrated management of water and soil. In particular, since the 1980s this approach has demonstrated, in various European and world

M.L. Scaduto, *River Contracts and Integrated Water Management in Europe*,
UNIPA Springer Series, DOI 10.1007/978-3-319-42628-0_1

contexts, its ability to address the related issues and support the dialogue and integration between public and private stakeholders.

The evaluation of the effectiveness of the RC, as well as the analysis of its horizontal and vertical relationships with urban and territorial planning, require a research methodology properly oriented to a comparative approach.

In this research, such methodology was applied to the national contexts of France and Italy, in order to analyze and better understand the European scenario.

1.2 Theoretical Framework of the Research

In the scientific, technical and politico-institutional fields, the need to focus and reflect on coherent and integrated water management at the river basin scale, is widely recognized (Burton 2002; GWP-RIOB 2009; Choukr-Allah et al. 2012; UNEP 2012). In this perspective, the importance of the social and the political dimensions is increasingly evident, as prerequisites for the achievement of sustainable development (Johnson et al. 2001; Teodosiu et al. 2003; Kemper et al. 2007a), also in the light of the awareness that water and territory are inseparable resources.

According to experts, integrated water management «should be managed based on river basins, not only on administrative boundaries» (Rahaman and Varis 2005, 19). In fact, in most cases the river basin represents the optimal spatial unit to structure and implement appropriate policies and procedural instruments (EC 2000; Teodosiu et al. 2003). In this sense, while being a geographical unit strictly connected to hydrogeological dynamics and functioning, the river basin has progressively become «a political and ideological construct» (Molle 2006, 23), capable to better support a shared management of water resources. To say it with Jaspers (2003, 81), «water necessarily has to be managed on hydrological boundaries, because water simply tends to flow down and it does not stop at the boundary of the district or region».

Although the history of hydrographical studies originates in Mesopotamic civilization, it was only in the second half of eighteenth century France, with the *Essai de géographie physique* of Philippe Buache (1752), that the river basin was first explicitly defined as a natural territorial unit, hence taken as reference for the establishment of administrative *départements* in 1789.

However, only in the beginning of the twentieth century the river basin actually became in different national contexts the acknowledged area to target interventions of economic and technical planning, such as in Spain with the *Confederaciones Hidrográficas* (1926), in the United States and in the former Soviet Union (Embid 2003; Molle 2006). In the European scenario, between the 1960s and the 1970s France and the United Kingdom led the way with two major initiatives (Barraqué 1995; Lasserre and Brun 2007). In 1964, the first *Loi sur l'eau* was promulgated in France and the six *Agences financières de bassin* were established, in order to redistribute, at the level of each river basin, the functions of integrated water management and five-year planning so as to achieve river quality objectives. Likewise, in 1974 the United Kingdom established the ten *Regional Water Authorities*, in charge of improving the quality of water resources, at the river basin scale.

In this scenario, *River Basin Management* (RBM) arised as a paradigm of management and planning (Teclaff 1996; Burton 2003). The RBM is the result of a long-lived and complex process, which started in different geographical contexts, steadily evolving for different purposes, «at the endless search for elusive governance units that would unite nature and societies» (Molle 2006, 24). In the RBM perspective, four priorities was identified: (I) overcoming issues related to institutional and administrative boundaries; (II) cooperation in fostering up-stream and down-stream relations; (III) stakeholder participation; (IV) appropriate decentralization of institutional competences.

The principle of River Basin Management did not find its way onto the international agenda until the early 1990s (Burton 2003). In fact, although in 1977 the *United Nation Conference in Mar del Plata* had identified the integrated management of water resources as a pillar of the *Mar del Plata Action Plan,* during the 1980s this strategic challenge disappeared from the international political debate (Rahaman and Varis 2005; Molle 2006).

At the beginning of the following decade, thanks to the efforts of various organizations and on the basis of considerations emerged at, and disseminated through a series of conferences, a new awareness began to spread on the international scene with respect to water management issues that «have become multi-dimensional, multi-sectoral and multi-regional, and filled with multi-interests, multi-agendas and multi-causes, and which can be resolved only through a proper multi-institutional and multi-stakeholder coordination» (Biswas 2004a, 249). In fact, during the 1992 *International Conference on Water and Environment* held in Dublin, and precisely within the so-called *Dublin Principles*, hydrographic basin-based integrated management was analyzed through a new holistic approach including forms of governance and stakeholder participatory actions, so as to take their effects into account from both economic and social perspectives (Burton 2002; Molle 2006).

In 2000, the *Second World Water Forum*, held in The Hague, universally acknowledged the river basin as the most suitable geographical unit for the management of water resources, besides being a vehicle for promoting territorial cooperation between stakeholders (Burton 2002).

As of the year 2000, the paradigm of Integrated Water Resources Management has also emerged. It was initially sponsored by the *Global Water Partnership*, by the corresponding *Global Water Forums* and through major international initiatives promoted by the *United Nations Program for Development*, *UN-Water*, *World Bank*, *World Water Council*, and others (UNEP 2012).

Within the IWRM theoretical and procedural framework, a special subset of specific actions was developed, namely the one dubbed *Integrated River Basin Management* (IRBM), and oriented to the management of all water resources, both in surface and subsurface. Particular attention was addressed to quality issues and participatory processes, to enhance the integration of all social, economic and environmental components (Jaspers 2003; Turton et al. 2007; Hamdy and Choukr-Allah 2012; Schnepf and Lutter 2012). Therefore, IRBM is based on the acknowledgement of two key concepts: (I) all components of the water cycle must be managed within a coherent territorial and management unit; (II) all stakeholders should be involved in decision-making and management processes.

According to Molle (2006), the emergence of IWRM and river basin as its reference unit is related to the confluence of four strands of thought: (I) the eco-systemic approach as a strategy for the integrated management of soil, water and biological resources; (II) the increasing weight of economic aspects in water management; (III) the need to take in account the up-stream and down-stream relations; (IV) the importance of stakeholder participation in line with the broader principle of subsidiarity.

The great potential and the degree of theoretical and procedural evolution that characterize the complex framework hitherto described must come to terms with a host of challenges in different territorial contexts, both internationally and at the local level. In fact, the intrinsic characteristics of the water resource make its planning and management two very complex tasks (Biswas 2004a). Although the international community has a keen awareness of the issues relating to water management, the gap between theoretical aspects and practical applications remains very wide nonetheless, also because issues and solutions related to IWRM local implementation might not be readily adaptable to the all the different contexts (Biswas et al. 2005; Rahaman and Varis 2005; Kemper et al. 2007b; Rodríguez-Clemente and Hidalgo 2012; Mitchell 2015).

Hence, these management challenges are often linked to (I) qualitative and quantitative aspects of water resources, (II) inherent complexity of management practices, (III) the level of specific expertise of the overseeing institutions, (IV) availability of adequate funding and, finally, (V) local environmental and socio-political conditions that profoundly influence water resource planning (Biswas 2004b). Specifically, with its emphasis on the need to deal with surface and underground water resources, as a whole, from the technical, political, economic and social points of view, the IWRM implies a double level of integration: (I) horizontally, between resources, uses and stakeholders, and (II) vertically, between different management scales (Charnay 2011). This entails participation, decentralization of management functions and innovative transnational and multi-disciplinary approaches (Burton 2002, 2003; GWP-RIOB 2009). These aspects make it blatantly explicit that the guiding principles of IWRM are markedly ambitious, rendering the array of interrelated objectives a fundamental «challenge for the current century» (Molle 2006, 22).

At the Second World Water Forum (2000), the RC was identified as an instrument that allows to adopt a system of rules in which the criteria of public interest, economic performance, social value and environmental sustainability are equally effective in finding solutions for the redevelopment of a river basin.

Due to the importance of the river basin in the management of water resources and notwithstanding the main practical limitations mentioned above, the RC can provide a complementary tool to facilitate regulation and integrated management of the river basin territory (Brun 2014). In fact, it involves several orders of interrelationships: longitudinal, between areas up-stream and down-stream to the basin; transversal, between the various socio-economic actors, and scientific, between different disciplines (geomorphology, biology, chemistry, economics, urban and regional planning, sociology, etc.) (Mostert et al. 1999). Consequently, the RC provides concrete evidence that governance of water resource is actually possible (Rosillon and Vander Borght 2001).

At the river basin scale, among the various obstacles and limitations to the implementation of integrated water management, the main one is precisely represented by achieving effective integration between the various administrative levels and actors involved (Lasserre and Brun 2007; Mitchell 2015), once what is meant by effective integration has been duly clarified (Affeltranger and Lasserre 2003; Moss 2003). Blonquist (2008), for example, highlights the complexity and difficulties arising from the great variety of interconnections between water resources (rivers, lakes, aquifers, groundwater, wetlands, etc.), communities and activities. Last but not least, in many cases these issues are clearly due to the mismatch between administrative boundaries and hydrographic basins, which is the single most-limiting factor facing an effective implementation of the management paradigm based on natural units.

Notwithstanding the many critical views of a number of authors, especially with respect to the actual scope of the frameworks hitherto described (Biswas 2004a; Molle 2006; Butterworth et al. 2010), an ongoing widespread debate has been focusing on the systemic themes of over-exploitation of aquifers, the impact of diffuse pollution, the importance of more rational use of water resources and also the need for participatory processes (UNEP 2012).

Also by means of the objectives and results of the *European Community Sixth Framework Programme for Research and Technological Development*, the reflections on the different declensions of coordinated water resource management culminated in the issuing of the European *Water Framework Directive* (WFD), in force as of 2000 (EC 2000). This directive is the product of thirty-year effort of the European Union in terms of water resource policies, provides a host of innovations and calls for member States to achieve, by 2015, a very ambitious goal: a clean bill of health for all surface, underground and coastal waters. The *conditio sine qua non* for the achievement of this goal is the implementation of coordinated planning processes capable of ensuring the participation of all stakeholders of each hydrographic district (Pahl-Wostl et al. 2008; Richter et al. 2013).

In light of the close correlations between the IWRM paradigm and the WFD (Teodosiu et al. 2003), even for the latter EC directive there are several barriers to the application of its recommendations, since in many cases the principle of subsidiarity may be contradicted especially for very large river basins (Rahaman et al. 2004; Molle 2006).

In this scenario, the French context provides a solid reference model, especially regarding the integration between the different levels of regulation and management of water resources (Richard et al. 2010). In 1992, the second *Loi sur l'Eau* introduced a hierarchy of regulatory instruments, in decreasing order going from the

Schéma Directeur d'Aménagement et Gestion des Eaux (SDAGE) for the scale of the main hydrographical basin, to the *Schéma d'Aménagement et Gestion des Eaux* (SAGE) for sub-basins, up to the various declensions of RC tailored to the specific functional and management needs of each local context (Brun 2014).

In France, the RC, in the form of the *contrat de rivière* (*CdR*), is an action plan supporting water management, according to which a moral commitment is formalized between its public and private co-signatories. The emergence and dissemination of these contractual agreements have characterized the evolution of water resource management in France. That process started after the mid 1960s, facilitating the passage from vertical, top-down public actions to horizontal and polycentric systems based on mutual cooperation of different actors (Brun 2010).

In France, *CdR* are part of the environmental agreements, representing a commitment on behalf of the co-signatories of a joint project, (Brun 2010, 2014). Moreover, in line with the Principles of Dublin they operatively aim to achieve the objectives of integrated water management at the river basin scale (Brun and Lasserre 2006). Therefore, as contractual deeds, they represent voluntary agreements between public and often private actors that, each within the framework of his own specific responsibilities, resolve to pursue a common project aimed at harmonizing the multiple uses and functions of waterways and water resources of an entire river basin (Bobbio 2006).

In the French context, the first experiences of *contrats de rivière* began in the early 1980s on the initiative of the Ministry of the Environment, with the signature of the first agreement at Thur in 1983. Since 1992, with the proclamation of the second *Loi sur l'eau*, those agreements have been recognized as the means of implementation of the *Schémas Aménagement et de Gestion des Eaux* (Lascoumes and Le Bourhis 1998). This current scenario totals 269 applications of *CdR* at different stages of implementation, among which 29 are cross-border initiatives involving areas of Spain, Belgium and Switzerland (http://www.gesteau.eaufrance.fr/).

In these countries the RC have been promoted on the basis of the pioneer experiences in France, as is also the case with Luxembourg. More in general, their diffusion in Europe has been fostered by EC stances that increasingly recognize a prominent role to contractual tools, highlighting the importance of dialogue between different actors.

Since 2003, even in Italy RC are increasingly being implemented nationwide, in the form of the *contratto di fiume* (*CdF*), and since 2008 was established the *National Board on River Contracts*. In this interdisciplinary workgroup both public administrations and local authorities, as well as technical experts, researchers and stakeholder associations come together for the promotion and exchange of best practices. Since 2008, ten national technical workshop have been organized and, at the fift held in Milan in 2010, the *National Charter of River Contracts* was presented as the first official reference document for the implementation of this kind of agreement in Italy (Bastiani 2011).

This series of initiatives constitutes an actual nationwide movement in which RC are seen as instruments for developing negotiated action plans aiming to re-qualify river basins, yet profoundly intertwined with a variety of territorial planning processes (Magnaghi 2008, 2011), thus facilitating the transition from management plans on the basin scale to those tailored to sub-basins.

In this perspective, RC may contribute to developing also in Italy new integrated forms of urban and regional planning and, therefore, represent an innovative instrument of territorial governance. Indeed, they are becoming effective tools for identifying shared strategies, actions and rules for the horizontal and vertical integration of policies, programs, action plans, for the purposes of fostering the participation of local communities and re-qualifying each river basin, even from socio-economic, landscape and environmental standpoints (Bastiani 2011).

Another key aspect of the RC paradigm is the voluntary participation of those stakeholders seeking to define and implement integrated and shared local water management actions. In this sense, these contractual agreements may help overcome the traditional mind-set within the specific sector of water and environmental resource management (Magnaghi 2008; Rosillon and Lobet 2008).

However, there are still many open issues with regard to the effectiveness of the RC in promoting the integration of policies concerning river areas, as well as with regard to its practical integration with other territorial action plans already in force, as highlighted by the scientific community (Brun 2014).

Within this complex theoretical and applicative framework, the strong interest of public administrations, scholars and researches, and local communities for the innovative RC paradigm requires a deeper understanding of its scope in terms of regulation, river basin requalification and effective integration into sectorial policies, and with urban and territorial planning.

1.3 Methodological Approach to River Contract Analysis

In order to provide an analytical framework for evaluating RC and their horizontal and vertical relationships with urban and regional planning, an appropriate methodological approach has been defined by focusing on the relationship between the research topic and the specific access keys necessary to achieve a deeper knowledge of the matter.

The complexity of the theme and its particular actuation among the various national contexts, initially prompted an analytical investigation spanning the whole European scenario, in order to identify the most paradigmatic case studies, so as to make critical comparisons, and highlight any valuable knowledge and interpretative aspects. Therefore, the focus was primarily on (I) the nature of the RC paradigm and its local declensions; (II) the different implementation modalities with respect to the various morphological, physical, institutional, social and economic contexts investigated; finally, (III) the evaluation of effectiveness and portability of models across different European contexts.

The method of empirical research was applied in a circular, bidirectional process composed of five phases and moving forward and backward with respect to each one phase (Agodi 1995). Specifically, these steps correspond to the (I) identification of the primary theme and definition of topics and questions, (II) design of the research model, (III) data organization and modelling, (IV) data coding and analysis; (V) interpretation of results (Fig. 1.1).

The methodological approach was thus subdivided into two fundamental interacting areas: the first regarding the general organization of the research and its methodological basis; the second consisting in the definition of case studies, as a specific application of the general method. In each area, an integration of both qualitative and quantitative methodological approaches, procedures and techniques was sought, as are widely used in the field of sociological research (Delli Zotti 1996).

Once defined the analytical and knowledge framework, the actual comparative investigation phase commenced on the four selected case studies related to France and Italy, particularly with respect to national and regional regulatory frameworks of reference, and to their relationships with planning experiences and instruments.

The comparative approach was chosen for its unquestioned validity, as is widely recognized in the literature (Hantrais 1995; Delli Zotti 1996; Vigour 2005). Some authors identify this as a fundamental method (Collier 1993) and an inevitable instrument in the researcher's toolbox (Sartori 1994). In fact, it permits discernment of similarities and differences between identical and/or different phenomena, with a diachronic vision for each moment and context (Marradi 1985). Moreover, the comparative method can be applied both through a single analytical technique, as well as through a battery of techniques, thus fostering a multi-faceted perspective on the analyzed phenomena (Delli Zotti 1996).

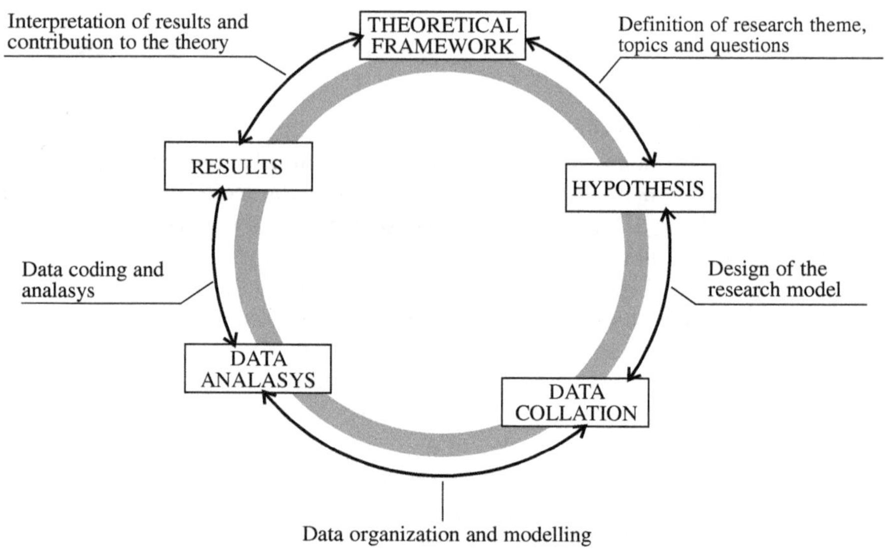

Fig. 1.1 Circular bidirectional process applied in the research

Specifically, the comparative method finds full application even in the analysis of public policies, in which the need for international comparisons is increasingly more evident (Sartori 1994; Hyman 1998; Hassenteufel 2005; Vigour 2005), especially in Europe, in light of the heightened interdependencies of political systems as a result of the growing tide of *européanisation* (Barbier 2005; Hassenteufel 2005). With these premises, the comparative approach was used in the present research in order to prove or invalidate the general hypotheses underpinning the investigation and, in this sense, it was applied from both descriptive and explanatory perspectives.

Specifically, the choice to undertake a comparative analysis between France and Italy derived from the objective to understand theoretical, regulatory, institutional and technical differences in the application of the RC paradigm in the two national contexts. Therefore, this analysis aimed to highlight a number of variables across the two selected national contexts, in order to assess their weight and effect with respect to the research topics. Due to these reasons, the investigation was mainly based on direct dialogue and in-presence discussions with Italian and French institutions, stakeholders and experts, according to the survey technique labeled "not at distance" (Hantrais 1995; Seiler 2004; Hassenteufel 2005).

In methodological terms, this study represents a binary comparison circumscribed to the two above mentioned countries that differ in geographic, socio-economic, historical and territorial characterizations, specifically chosen on the basis of the preliminary analysis of the investigation domain (Delli Zotti 1996). This international comparison obviously took into account the relative distance of the two national contexts from each other, with regard to the specific matter, both in space (synchronic comparison) and in time (diachronic comparison).

Along this methodological path a particular attention was paid to what Delli Zotti (1996, 159) defines the *danger of nominalism*. In this perspective, the actual role that the two very similar instruments of *CdR* and *CdF* play in their respective contexts, was extensively scrutinized, also observing various declensions of the RC paradigm in Europe.

Accordingly, the study was conducted through field data collection, analysis and comparison and, finally, via a specific design and elaboration of an analytical matrix of selected case studies. With regard to the contents, the comparative study at the national scale was divided into four levels: (I) normative references, (II) contents and procedures, (III) actors and stakeholders playing key roles in the implementation of RC, and (IV) completed and ongoing experiences. These analytical levels served to delineate a clearer application panorama of this type of agreement.

This initial survey represented the essential starting point for the choice of case studies useful to verify the theoretical research hypotheses. Within the methodological process and particularly during the construction of knowledge bases, the case study-based approach permitted to analyze and understand the investigated phenomena and processes, from a holistic and multi-perspective outlook (Feagin et al. 1991; Yin 1994; Tellis 1997).

Although the typical limits of this methodological approach are well known, the case study method maintains some distinctive characteristics that make it very suitable for many types of investigation, also in combination with other methods (Yin 1994; Tellis 1997; Zaidah 2007).

Despite some disadvantages inherent to this approach, such as the risk of lacking rigor, varying degrees of generalizability of results and the bulk of documentation to process, nonetheless the case study-based research, on the one hand, satisfies the criteria of qualitative methods (description, understanding and explanation) and, on the other hand, also makes use of survey tools that are based on multiple and comparative analysis schemas (Hamel et al. 1993; Yin 1994; Stake 1995; Delli Zotti 1996; McDonough and McDonough 1997).

Specifically, on the basis of the methodological integration between qualitative and quantitative analyses, the following items emerged as principal aspects of the survey:

- exploratory and descriptive approaches to the continuous evolution and complexity of the phenomena and processes involved in the application of RC, and in their integration with urban-territorial and river basin planning instruments;
- unit of analysis, defined by means of a multiple-case approach, in order to provide solid bases to the comparative analysis;
- choice of case studies, made through specific selection criteria, such as (I) relevance to the research questions (Ricolfi 1997); (II) representativeness of each case; (III) innovation in integration with management and planning tools; (IV) dimensional criteria of the river basin; (V) geographical localization criteria; finally, (VI) availability of and access to data and documentation for each case (Mason 1996);
- interpretative and comparative matrix of the case studies for the systematization of the collected data; this tool served to precisely orient the subsequent more comprehensive comparative interpretation that outlined the conclusions of the research.

The application of the described method was integrated by critical examination of the sectorial scientific literature, summary reports on the thematic boards, agreement protocols, minutes of meetings, action plans, RC, etc.

Taking into account the critical elements of the specific matter of the present research, and the open issues still to be better analyzed, four case studies were selected as the most representative, in order to conduct the comparative study of French and Italian RC frameworks. In this perspective, the careful selection of case studies was oriented towards a cross-comparison of river basin management experiences carried out both in urbanized and in rural contexts, as will be seen in detail in Chap. 4.

The final objective of this case-study comparison was to identify and describe the theoretical, procedural and applicative elements to be potentially integrated into a river contract implementation model that could be transferred more easily to other European and world contexts, just with some adaptation to local geographical, hydrographic, institutional and socio-economical situations.

References

Affeltranger B, Lasserre F (2003) La gestion par bassin versant: du principe écologique à la contrainte politique – le cas du Mékong. VertigO - la revue électronique en sciences de l'environnement [En ligne] 4 (3)

Agodi MC (1995) Qualità e quantità: un falso problema e tanti equivoci. In: Cipolla C, de Lillo A (eds) Il sociologo e le sirene. La sfida dei metodi qualitativi, Franco Angeli, Milano, pp 106–135

Barbier J-C (2005) Remettre la comparaison internationale sur l'ouvrage et dans ses mots. In: Barbier J-C, Letablier M-T (eds) Politiques sociales. Enjeux méthodologiques et épistémologiques des comparaisons internationales, Peter Lang, Bruxelles, pp 17–43

Barraqué B (1995) Les Politiques de l'eau en Europe. La Découverte, Paris

Bastiani M (2011) Dalla valorizzazione degli ambiti fluviali ai contratti di fiume. In: Bastiani M (ed) Contratti di Fiume. Pianificazione strategica e partecipata dei bacini idrografici. Approcci, esperienze, casi studio. Dario Flaccovio Editore, Palermo, pp 3–30

Biswas AK (2004a) Integrated water resources management: a reassessment. Water Int 29(2): 248–256

Biswas AK (2004b) From Mar del Plata to Kyoto: a review of global water policy dialogues. Glob Environ Change Part A(14):81–88

Biswas AK, Varis O, Tortajada C (2005) Integrated water resources management in South and Southeast Asia. Oxford University Press, New Delhi

Blonquist W (2008) Relating people and the environment in river basins. Celebrazione 60° anniversario programma Fulbright USA-Italia, 29 aprile 2008, Torino

Bobbio L (2006) Le politiche contrattualizzate. In: Donolo C (ed) Il futuro delle politiche pubbliche. Paravia Bruno Mondadori Editori, Milano, pp 59–79

Brun A (2010) Les contrats de rivière en France: un outil de gestion concertée de la ressource en local. In: Schneier-Madanes G (ed) L'eau mondialisée. La gouvernance en question, Éditions La Découverte, Paris, pp 305–321

Brun A (2014) France's water policy: the interest and limits of river contracts. In: Schneier-Madanes G (ed) Globalized water: a question of governance. Springer, Dordrecht, pp 139–147

Brun A, Lasserre F (2006) Politiques de l'eau. Grands principes et réalités locales. Presses de l'Université du Québec, Québec

Burton J (2002) La gestion intégrée des ressources en eau par bassin au-delà de la rhétorique. In: Lasserre F, Descroix L (eds) Eaux et territoires. Tensions, coopérations et géopolitique de l'eau. Presses de l'Université du Québec, Québec, pp 189–207

Burton J (2003) Integrated Water Resources Management on a Basin Level. A training manual. UNESCO, Canada

Butterworth J, Warner J, Patrick M, Smits S, Batchelor C (2010) Finding practical approaches to integrate water resources management. Water Altern 3:68–81

Charnay B (2011) A system method for the assessment of integrated water resources management (IWRM) in mountain watershed areas: the case of the "Giffre" watershed (France). Environ Manage 48(1):189–197

Choukr-Allah R, Ragab R, Rodriguez-Clemente R (2012) Integrated water resources management in the Mediterranean region. Springer, Dordrecht

Collier D (1993) The comparative method. In: Finifter AW (ed) Political science: the state of the discipline. American Political Science Association, Washington

Delli Zotti G (1996) Il metodo comparato in sociologia. In: Gasparini A, Strassoldo R (eds) Tipi ideali e società. Franco Angeli, Milano, pp 151–178

Embid A (2003) The transfer from the Ebro basin to the Mediterranean basins as a decision of the 2001 national hydrological plan: the main problems posed. Int J Water Resour Dev 19: 399–411

European Commission (EC) (2000) Directive 2000/60/EC of the European Parliament and of the Council of 23 October 2000 establishing a framework for Community action in the field of water policy. Off J Eur Commun L 327(22) Dec. 2000

European Commission (EC) (2007) Directive 2007/60/EC of the European Parliament and of the Council of 23 October 2007 on the assessment and management of flood risks. Off J Eur Commun L 288(27) Nov. 2007

Feagin J, Orum A, Sjoberg G (1991) A case for case study. University of North Carolina Press, Chapel Hill

GWP (2000) Integrated water resources management. Technical Advisory Committee Background Paper n. 4. Global Water Partnership—Technical Advisory Committee, Stockholm

GWP-RIOB (2009) Manuel de Gestion Intégrée des Ressources en Eau par Bassin. Elanders, Suède

Hamdy A, Choukr-Allah R (2012) Effective water governance and how to achieve. In: Choukr-Allah R, Ragab R, Rodriguez-Clemente R (eds) Integrated water resources management in the Mediterranean region. Springer, Dordrecht, pp 267–284

Hamel J, Dufour S, Fortin D (1993) Case study methods. Sage Publications, Newbury Park

Hantrais L (1995) Comparative research methods. Social research update. University of Surrey, Guildford

Hassenteufel P (2005) De la comparaison internationale à la comparaison transnationale. Les déplacements de la construction d'objets comparatifs en matière de politiques publiques. Revue française de science politique 55:113–132

Hyman R (1998) Recherche sur les syndicats et comparaison internationale. Revue de l'Ires 28:43–61

Jaspers FGW (2003) Institutional arrangements for integrated river basin management. Water Policy 5:77–90

Johnson NL, Ravnborg HM, Westermann O, Probst K (2001) User participation in watershed management and research. Water Policy 3:507–520

Kemper KE, Blomquist W, Dinar A (2007a) Integrated river basin management through decentralization. Springer, Dordrecht

Kemper KE, Blomquist W, Dinar A (2007b) River basin management at the lowest appropriate level: when and why does it (not) work in practice? In: Kemper KE, Blomquist W, Dinar A (eds) Integrated river basin management through decentralization. Springer, Dordrecht, pp 3–15

Lascoumes P, Le Bourhis J-P (1998) Les politiques de l'eau: enjeux et problématiques. Regards sur l'actualité 241:33–41

Lasserre F, Brun A (2007) La gestion par bassin versant: un outil de résolution des conflits? Lex Electron 12(2):1–19

Magnaghi A (2008) I contratti di fiume: una lunga marcia verso nuove forme integrate di pianificazione territoriale. Notiziario dell'Archivio Osvaldo Piacentini 1:89–98

Magnaghi A (2011) Contratti di fiume e pianificazione: uno strumento innovativo per il governo del territorio. In: Bastiani M (ed) Contratti di Fiume. Pianificazione strategica e partecipata dei bacini idrografici. Approcci, esperienze, casi studio. Dario Flaccovio Editore, Palermo, pp 31–44

Marradi A (1985) Natura, forme e scopi della comparazione: un bilancio. In: Fisichella D (ed) Metodo scientifico e ricerca politica. La Nuova Italia Scientifica, Roma, pp 293–321

Mason J (1996) Qualitative researching. Sage, London

McDonough J, McDonough S (1997) Research methods for English language teachers. Arnold, London

Mitchell B (2015) Water risk management, governance, IWRM and implementation. In: Urbano FP (ed) Risk governance the articulation of hazard, politics and ecology. Springer, Dordrecht

Molle F (2006) Planning and managing water resources at the river-basin level: emergence and evolution of a concept (IWMI Comprehensive Assessment Research Report 16). International Water Management Institute, Colombo

Moss T (2003) Solving problems of 'Fit' at the expense of problems of 'Inter-play'? the spatial reorganisation of water management following the EU Water Framework Diective. Institute for Regional Development and Structural Planning, Erkner

Mostert E, Van Beek E, Bouman NWM, Hey E, Savenije HHG, Thissen WAH (1999) River basin management and planning. In: Mostert E (ed) IHP-V Technical Document in Hydrology, vol 31. pp 24–55

Pahl-Wostl C, Tabara D, Bouwen R, Craps M, Dewulf A, Mostert E, Ridder D, Taillieu T (2008) The importance of social learning and culture for sustainable water management. Ecol Econ 64:484–495

Rahaman MM, Varis O (2005) Integrated water resources management: evolution, prospects and future challenges. Sustain Sci Pract Policy 1(1):15–21

Rahaman MM, Varis O, Kajander T (2004) EU water framework directive vs. integrated water resources management: the seven mismatches. Water Resour Dev 20(4):565–575

Richard S, Bouleau G, Barone S (2010) Water governance in France: institutional framework, stakeholders, arrangements and process. In: Jacobi P, Sinisgali P (eds) Water governance and public policies in Latin America and Europe. Anna Blume, pp 137–178

Richter S, Völker J, Borchardt D, Mohaupt V (2013) The water framework directive as an approach for integrated water resources management: results from the experiences in Germany on implementation, and future perspectives. Environ Earth Sci 69:719–728

Ricolfi L (1997) La ricerca empirica nelle scienze sociali. Una tassonomia. Rassegna italiana di Sociologia XXXVI(3):339–418

Rodríguez-Clemente R, Hidalgo A (2012) Identifying risks, actions, and opportunities for a long-term water strategy in the Mediterranean. In: Choukr-Allah R, Ragab R, Rodriguez-Clemente R (eds) Integrated water resources management in the Mediterranean region. Springer, Dordrecht

Rosillon F, Lobet J (2008) Transboundary river contract Semois-Semoy between Belgium (Wallonia) and France. In: Meire P, Coenen M, Lombardo C, Robba M, Sacile R (eds) Integrated water management. Springer, Dordrecht

Rosillon F, Vander Borght P (2001) Mobilisation au fil de l'eau: dix années d'expérience de CdR en Wallonie. Annales de Mines, pp 39–54

Sartori G (1994) Bien comparer, mal comparer. Revue Internationale de Politique Comparée 1(1):19–36

Schnepf D, Lutter S (2012) Integrated participatory water resources management and water governance. In: Choukr-Allah R, Ragab R, Rodriguez-Clemente R (eds) Integrated water resources management in the Mediterranean region. Springer, Dordrecht

Seiler DL (2004) La méthode comparative en science politique. Armand Colin, Paris

Stake RE (1995) The art of case study research: perspective in practice. Sage, London

Teclaff LA (1996) Evolution of the river basin concept in national and international water law. Nat Resour J 36(2):359–391

Tellis W (1997) Application of a case study methodology. Qual Rep 3(3):1–17

Teodosiu C, Barjoveanu G, Teleman D (2003) Sustainable water resources management. 1. River basin management and the EC water framework directive. Environ Eng Manag J 2(4):377–394

Turton AR, Hattingh HJ, Maree GA, Roux DJ, Classen M, Strydom WF (2007) Governance as a trialogue: government–society–science in transition. Springer, Berlin

UNEP (2012) The UN-water status report on the application of integrated approaches to water resources management. United Nations Environment Program

Vigour C (2005) La comparaison en sciences sociales: pratiques et méthodes. La Découverte, Paris

Yin R (1994) Case study research: design and methods, 2nd edn. Sage Publishing, Thousand Oaks

Zaidah Z (2007) Case study as a research method. Jurnal Kemanusiaan 9:1–6

Chapter 2
River Contracts for Innovation in Territorial Governance

Abstract Emerged in France in 1980s as agreements oriented to the requalification of rivers and lakes, river contracts represent an important outcome of the decentralization process, developed in Europe in reply to the growing institutional fragmentation. In this scenario, rivers have become a dialectic arena in which the exploitation and management of water resources came to terms with environmental instances, and offered the breeding ground for concerted efforts between policy makers, stakeholders and communities. In this chapter, the analysis is focused on the river contract model and the related set of instruments capable of supporting concerted and participatory management processes both with respect to European and national policy frameworks, and the integration with urban an territorial planning.

2.1 Introduction

The river contracts (RC) originate in France in the early 1980s as mid- or long-term requalification programs for rivers, lakes, aquifers and river mouths, based on the consultation among stakeholders. Within the two ensuing decades they had become a pivotal tool in French policy regarding the integrated management of water resources, as well as a reference paradigm for the requalification of watercourses at the river basin scale (Brun and Marette 2003; Brun 2010a, 2010b, 2014).

In Europe, RC represent one of the outcomes of the decentralization process at the institutional and bargaining policy levels, launched by the European Community, as of the early 1980s, expressly in reply to the growing institutional fragmentation (Sancy 2008). The RC go under the heading of so-called voluntary agreements, originating and spreading in France and Germany in the early 1970s (Orts and Deketelaere 2001). Since by there, it was clear that this kind of instrument was capable to foster novel forms of dialogue and shared responsibility among public and private actors and, thus, to support new processes of local governance (Delmas and Terlaak 2001; Faure 2001; Hervé-Fournereau 2008).

M.L. Scaduto, *River Contracts and Integrated Water Management in Europe*,
UNIPA Springer Series, DOI 10.1007/978-3-319-42628-0_2

Some underlying reasons typical of environmental management policies have contributed to the wide dissemination of RC: (I) the marked complexity of environmental issues, (II) the need to combine the contributions of various disciplines and actors, (III) the crucial role of consultation and contractual practices, (IV) the ever-widening active participation in water management on the part of society.

These four components have highlighted the fact that rivers represent a dialectic arena, for interest groups conveying environmental demands and others seeking to exploit the diverse uses of water resources, and therefore a battleground at times laden with strife, but nevertheless a stage for reflection and concerted efforts between policy makers, stakeholders and more recent figure as the so-called *boundary workers* (Gailliard et al. 2014).

As said, the first definition of the RC paradigm was proposed during the Second World Water Forum (2000). Such definition represented an important breakthrough in terms of applicability with reference to the concept of integrated management of waters, as had already been defined in Dublin in 1992, in occasion of the *International Conference on Water and the Environment* (Solanes and Gonzalez-Villarreal 1999; Brun and Lasserre 2006; Brun 2010a). In fact, at the Second World Water Forum, RC was defined as an instrument that allows to adopt a system of rules in which the criteria of public interest, economic performance, social value and environmental sustainability are equally effective in finding solutions for the redevelopment of a river basin, also in line with the European Water Framework Directive.

Although laced with some ambiguity from a legal standpoint (Brun 2010a, 2010b), the RC model was characterized as a novel instrument capable of setting in motion a processes of concerted and participatory management of water resources at the local scale (Bobbio and Saroglia 2008), spurring up-stream and down-stream players to greater synergy, as well as facilitating the application of the Integrated Water Resources Management (IWRM) paradigm to the each hydrographic basin context (Hooper 2005).

The weight of RC is particularly evident in the processes of building balanced synergies between policy makers and water users, also thanks to the boundary workers in the role of mediators and, thus, as advocates of conflict resolution between the calls for territorial development and safeguards for natural resources (Dervieux 2005; Gailliard et al. 2014).

2.2 River Contract in European Water Policies

At international level, the processes regarding the integrated management of water resources have been given compelling boosts towards innovation on a number of occasions arising as of the 1970s. Among the latter, the most significant events include: the *United Nations Conference on the Human Environment*, held in Stockholm in 1972, on which occasion was specifically highlighted the growing concern for threats posed to global water resources (UNEP 1972; Aubin and Varone

2001; Molle 2006); the first *UNESCO International Conference* of Mar de la Plata of 1977, where the importance of the Integrated Water Resources Management strategy for the resolution of conflicts relative to the different uses of water resources, was first acknowledged (Jeffrey and Gearey 2006); the *International Conference on Water and Environmental Issues for the 21st century*, held in Dublin in January 1992, that defined the guiding principles for actions at the local, national and international levels regarding environmental topics and water policies, including the integrated approach to water management and the need for participation of all stakeholders (Giordano and Wolf 2003; Teodosiu et al. 2003; Rahaman et al. 2004; Rahaman and Varis 2005).

Subsequent events marked other milestones in the long path to the affirmation of the IWRM paradigm. In 1992, in Rio de Janeiro, the *United Conference on Environment and Development* led to formulation of *Chapter 18 of Agenda 21*, entirely dedicated to the protection and integrated management of water resources, also relying on vital contributions from local communities in terms of information, awareness and participation. Since 1997, the *World Water Forums* have aimed at defining a broader and more global vision of economic issues and of participatory processes. In particular, the *Second World Water Forum*, held in The Hague in 2000, acknowledged the social, cultural and ecological values of water resources and singled out the IWRM as the only effective approach in the management of water resources, capitalizing on the results of previous initiatives (Shen and Varis 2000; Biswas 2004a; Rahaman and Varis 2005). At the same forum, the paradigm of the RC was also defined and acknowledged for the first time at the international level, as a viable instrument of integrated water management and territorial sustainable development.

Of course, this cursory overview of events should be integrated with a list of many other conferences and workshops of international level that likewise addressed the themes of management, conservation and consumption of water resources (Biswas 2004b; Rahaman et al. 2004; Rosillon and Lebeau 2010).

In Europe, the year 2000 represents a milestone also with regard to the definition of European Community and national water policies. In fact, the adoption of Directive 2000/60/EC *Water Framework Directive* (WFD) by the Council and the European Parliament, ratified the first European unitary framework of water resources integrated management, as the primary result of an articulate path characterized by a long sequence of EC programs and directives beginning in the early 1970s.

After a first *wave* of EC directives (1975–1990), aimed at protecting the quality of ground and surface waters depending on the specific local utilizations (Kaika 2003) and, after a second season of community legislation (1991–1999), mainly focused on environmental protection and control of emission levels (Aubin and Varone 2001; Kaika and Page 2003a, b), the WFD formalized the intent of the European Community to innovate its water policy through multi-sectoral approaches. In doing so, for the first time the EC defined a common and integrated framework for the management and protection of inland surface waters ground waters, transitional waters and coastal waters, thus laying the groundwork for the implementation of territorial governance and participatory and inclusive processes.

The WFD represents, therefore, the key document for reforming EC legislation on the matter (Kaczmarek 2003; Carter 2007). The promulgation of that directive is «a response to recent economic, political and social changes related to water management, including the shift from government to governance, the liberalization of water markets and the emergence of a new set of institutions, actors, etc., and their respective relations» (Kaika 2003, 299).

Integration constitutes a core concept in the context of the WFD and concerns the interrelation between different aspects such as environmental objectives, water resources at the river basin scale, the different uses, functions and values of waters, the competences and disciplines involved in water resource management, regulatory frameworks and EU, national and local legislations, local communities, the different decision-making levels. In addition, the WFD promotes the integration of different systems of water management among Member States.

Directive 2000/60/EC introduced seven innovative points which make reference to (I) the coordination of policies and strategies for water management, (II) the organization of water management based on river basins and not only on administrative boundaries, (III) the introduction of a combined approach to emission control and environmental quality standards, (IV) the introduction of quantitative criteria in the environmental protection action planning (V) the redefinition of *good water status* and the list of substances hazardous to health, (VI) the introduction of full cost pricing and environmental cost recovery into water pricing, and (VII) the improvement of involvement and participation of local communities (Kaika and Page 2003b).

The objectives at which the WFD aims are therefore to:

- provide an integrated water management system based on river basins rather than on political or administrative borders (art. 3);
- set environmental requirements for water quality protection and to achieve good water status of rivers, lakes, coastal waters and groundwater (art. 4);
- introduce a combined approach to emission controls and groundwater protection (art. 10);
- encourage the sustainable use of water resources (art. 5, 7 and 9);
- contribute to mitigating the effects of floods and droughts (art. 1);
- ensure the involvement and active participation of all the interested parties (art. 14).

To achieve good water status by 2015, the WFD defined specific milestones, including the completion and publication of second-generation river basin plans all within the same 2015 deadline.

In this EU policy scenario the long evolutionary process of RC was embedded, matching the 7 points of innovation introduced by the WFD. In particular, in relation to the directive transposition by Member States, since 2003, and its implementation at the local level, RC have become part of the set of particularly useful operational tools, capable to contribute to the EU water policies objectives achievement.

The transposition of the WFD by Member States has indeed triggered a multiplier effect, setting off a cascade of many and varied experiences of integrated water management, in which the RC has represented both an innovative and effective procedure. In particular, the paradigm shift from government to governance of river basin districts, introduced by Directive 2000/60/EC, draws attention to the actual difficulties regarding resource management no longer confined to mere administrative boundaries, whether local or national (Kaika 2003). It is precisely the need for novel forms of governance in the practical implementation of the WFD that requires more flexible programming and planning tools capable of being modeled according to the individual regional hydrographic or trans-boundary realities, and the corresponding geo-political and socio-economic contexts. Specifically, the transformations in European political orientation, commencing in the early 1990s, result in models of governance that have gradually determined a shift from top–down, centrally imposed policies to negotiated and concerted forms of agreement (Moss 2004).

RC, therefore, are having to deal with the renovated scenario determined by the WFD and the consequently new arrangement of the European hydrography, subdivided into river basin districts, comprising river basins and sub-basins. In this context, the introduction of this new territorial scale of reference emphasized the inevitable mismatch between hydrographic and political-administrative units, generating doubts and reflections concerning the actual capabilities of the relevant institutions to fulfill their responsibilities, create synergies between the public and private sectors, develop new scientific, technical and managerial skills, interact with the different institutional, economic and social networks involved (Moss 2004). In view of these critical elements, the negotiated approach inherent in RC has represented a valid operational practice to integrate the various demands of territorial players, so as to define a clear framework describing each actor responsibilities and competences, including those of local communities, on the one hand, and to mold the physiographic identity of each river basin while strengthening its unity, at both administrative and planning level, on the other.

The WFD, in this sense, has given new impulse to the spread and evolution of RC by pointing out the prospects for cooperation oriented towards the coordination between different institutional and administrative levels, together with the participation of local populations, all based on a multi-sectoral and multi-disciplinary approach to the management of water resources, moreover in line with the perspective of European legislators. Directive 2000/60/EC has indeed vigorously contributed to the revitalization of the lengthy evolutionary process of RC beginning in France in the 1980s and then extended from the 90's on to Belgium, Luxembourg, Spain, Switzerland, Italy, the Netherlands, Germany, England and Greece, whereas outside of Europe, to Quebec, Burkina Faso, Bolivia and Chile (Rosillon and Vander Borght 2001; Bobbio and Saroglia 2008; Brun 2010a; Bastiani 2011) (Fig. 2.1).

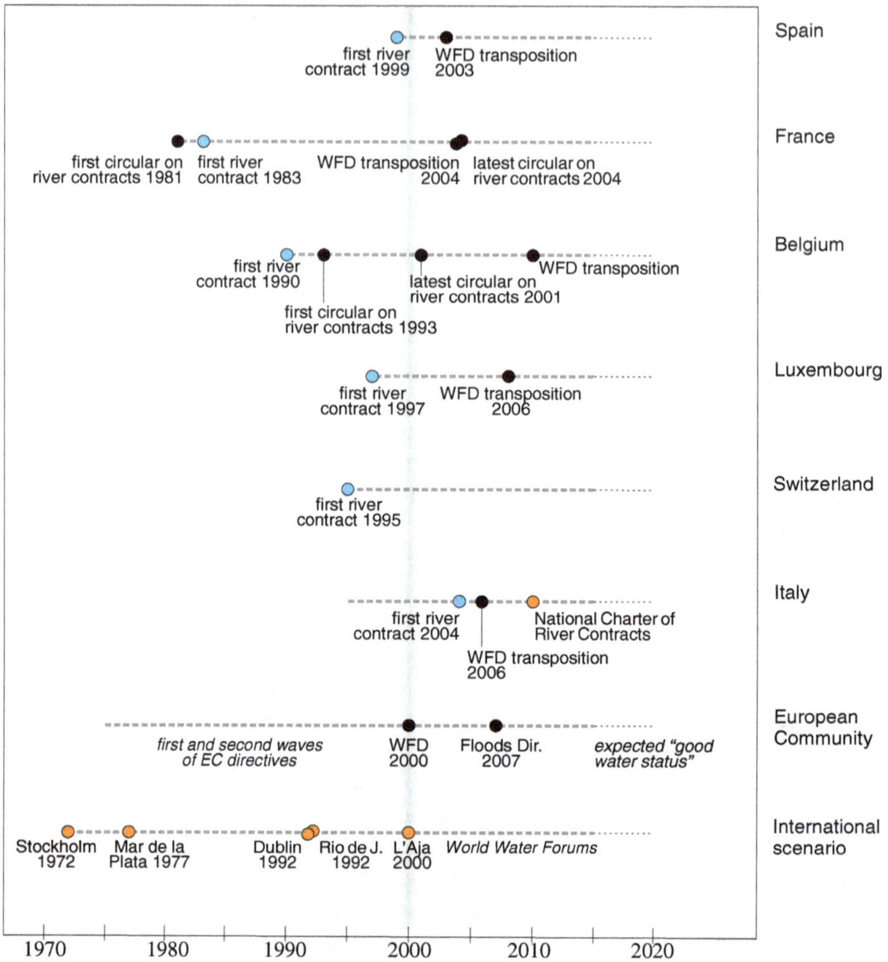

Fig. 2.1 Timeline of primary world water conferences and forums (*orange circles*), European and national legal milestones (*black circles*), and first national RC experiences (*blue circles*)

The projects already underway in several Member States at the time of enactment of the WFD, give important testimony of such long journey and, at the same time, help to delineate the current evolutionary stage of RC, which are increasingly being fully integrated into the normative and regulatory framework, as defined by the European Community as of the year 2000. The highlights of the various cases in which RC have been adopted within Europe concur in defining a more comprehensive framework of such negotiation instruments, within EC water policies, and their implementation at the level of individual Member States.

For example, in the case of Belgium, the Walloon Region, via the Circular issued March 20, 2001 and the subsequent Government Decree of November 13,

2008, sanctioned the role of RC in the implementation of management plans for river basin districts and, in compliance with Directive 2000/60/EC, formalized the requirement for these contracts to be revised accordingly so as to accommodate eventual sub-basins belonging to those river basin districts.

In contrast, the experiences occurring in Spain as of the 1990s particularly privileged coordination and cooperation aspects between Member States, as underscored by Directive 2000/60/EC, with regard to cross-border contexts involving France. These cross-border experiences, occasioned by the first contracts, led the *Confederación Hidrográfica del Ebro* to commission the development of a pilot project for the *Matarraña* river basin, in 2009, thus launching the first experimentation run of a RC conducted entirely on Spanish soil and, as such, oriented towards an integrated water management based on cooperation and coordination between the different administrative levels, as well as on the participation of local communities (Campos et al. 2011; Monge and Presa 2011).

RC implemented in Switzerland, in the cantons of Geneva and Jura, resemble the Spanish contexts, in that as far as their link to European policies is concerned, the salient element consists in the cross-border cooperation aspect with France, in whose territory can be found the headwaters of all major waterways running through Switzerland.

With regard to France and Italy, the experiences undertaken in the respective contractual contexts provide just as much testimony highlighting the impulse towards innovation and the diffusion of RC, occasioned by the WFD. To provide a more comprehensive account of the evolving role of RC, in particular regarding the implementation of EC water policies, as well as to furnish the specifics and an overall view of the general framework, the French and Italian case studies are illustrated in greater detail in Chap. 3.

2.3 River Contract in Integrated Management of Hydrographic Basins

Among the European Member States in which RC have been introduced, no univocal definition of them exists within their legal and regulatory frameworks. From a legal standpoint, these voluntary contractual agreements formalize moral and operational obligations of the co-signatories towards agreed plans of water resource management at the basin scale, including the technical and financial provisions and, thus, the actions programs required to achieve the joint objectives (Brun 2010a, 2010b, 2014; Allain 2004; Billet 2008).

In France the RC is defined as a joint technical and financial agreement, varying in duration from five to ten years, normally stipulated between the State, the *Régions*, the *Départements*, the *Agences de l'Eau*, local communities and other stakeholders of a same hydrographic basin (Brun 2014).

As for Belgium, it is interesting to note how the definition of such agreements is more oriented towards strengthening the dialogue and coordination among the various actors interested in building and sharing strategies, and programs of integrated actions, for the requalification, protection and development improvement of water resources of a hydrographic basin (Rosillon and Vander Borght 2001).

In Luxembourg the approach parallels that of the Belgian context with agreements negotiated between public and private parties. These RC do not have any predefined constraint on duration and are, in any case, characterized by particular regard to the sensitization and active participation of local communities (OECD 2010).

In Switzerland, RC take on the form of actual technical and financial agreements, whereby each signatory actor defines, with the other ones, objectives and specific actions, with particular regard to environmental recovery and the revitalization of watercourses, flood risk control and water resource management (OECD 2007).

In Spain, these contractual implements are defined by arrangements subscribed both by public institutions and private parties, and likewise to other European contexts they are capable to foster processes of public participation in managing water-related and environmental issues (Monge 2015).

In Italy, RC are characterized as continuous, multi-scalar processes of negotiated and participatory planning, geared to the containment of the environmental degradation of hydrographic landscapes and to the requalification of areas of basins and sub-basins (Bastiani 2011).

The analysis of the experiences carried out in various national contexts highlights that, aside from the local declensions of RC, there is a common need to involve water users, thus giving them a sense of responsibility, and to harmonize the often opposing key objectives of elevating the quality standards of water resources, and fostering local development.

In any cases, the various approaches to RC relate to the five common components deemed necessary for the integrated management of water resources, namely (I) a single reference unit in terms of hydrographic basins or sub-basins; (II) knowledge of water resources and the environmental, social and economic aspects correlated to their diverse uses; (III) voluntary acceptance of contractual instruments; (IV) coordination among all territorial actors; (V) participatory processes, with particular regard to involvement on the part of local communities.

In the European scenario, the capacity of RC to operate in integrated and cross-sectoral management of water resources is clearly emerging. In this sense, it represents a new versatile form of transposition and implementation of European and national water policies to local contexts, above all thanks to its multi-criteria scope and effectual sustainability that characterize many of these contractual agreements. Recurrence to RC is therefore warranted in those cases requiring interventions that target multiple structural causes of river degradation, and where the intent is to raise the quality of surface and ground waters, prevent and control hydrogeological risks and floods, requalify and develop fluvial and peri-fluvial zones, promote economic activities within river basins, as well as to assure adequate levels of information, education, sense of both individual and collective responsibility, and active involvement on the part of local communities.

In the light of the previous observations, it is possible to delineate in greater detail the role that RC may assume in the integrated management of water resources.

The reference to the territorial unit of the hydrographic basin represents a stimulus to simultaneously overcome limitations of two levels of institutional and management heterogeneity. The first level concerns the local administrative units that can identify in the RC the most expedient occasion to address management aspects linked to their institutional competencies and responsibilities, in a shared and integrated way. The second heterogeneity level is regarding the hierarchical relations and different water management competencies pertaining to many institutions, ranging from state to local government bodies.

In this perspective, municipalities above all must partake within new *water territories* and widen their oftentimes parochial views on water resource management, in favor of measures that are more aptly modeled onto systems with constituent up-stream and down-stream interconnections. Therefore, by way of agreed courses of action and programs, municipalities and territorial stakeholders can achieve an increased awareness of the extent of their interdependences and a deeper understanding of the need for a substantial *river basin solidarity*. In this direction, they can achieve the critical mass required to effectively transpose EU and national water policies to the local level and more readily gain access to European and local funding (Brun 2014).

In terms of knowledge of water resources, and of the environmental, social and economic aspects related to their different uses, RC can represent significant opportunities for public administrators to involve users even in the groundwork, starting from the preliminary fact-finding phase. The opportunity to collaboratively assemble a common asset of data and knowledge makes for a more insightful and comprehensive depiction of the issues and environmental and eco-systemic characteristics of water resources, as well as of water use and optimal management. The above constitutes a further incentive for launching innovative processes, through RC implementation, thanks to the underpinning approach based on multiple criteria applied to hydrographic territorial analysis and water resources integrated management.

The voluntary feature of these agreements, moreover, favors the recruitment of a host of assorted parties, even in different phases along the implementation pathway. All the more, even though they normally arise from voluntary, public or private initiatives, nevertheless they are capable to activate more flexible and effective forms of governance of water territories. In particular, RC constitute valid tools to deal with the complexities of local administrative and political scenarios, and their relative geographical contexts, whereas the higher-level normative frameworks often cannot be as readily modulated so as to generate sustainable actions with respect to local realities.

RC accrue further backing to the integrated management of water resources due to the interactions and interdependencies that, thanks to them, may be forged amongst actors. In particular, the coordination among all territorial actors, together with the participatory and collaborative approach, as well as the active involvement of local communities, represent important occasions for interaction and synergy finalized to the development of new forms of territorial governance (Tippet et al. 2005; Enserink et al. 2007; Bobbio and Saroglia 2008; Gailliard et al. 2014). In this perspective and from an institutional outlook, RC mark the transition from centralized and hierarchical forms of management to models that are instead decentralized and negotiated between public and private actors. The potential of these contractual tools finds its expression in relation to the identification and mobilization of active social and economic networks that can in turn be incorporated into the water governance process.

2.4 River Contract in Urban and Territorial Planning

At an international level, there is an increasingly recognized need for holistic and multi-sectoral approaches to urban and territorial planning. At the same time, the multi-sectoral perspective should favor synergies between different disciplinary approaches and the contents of the various plans and programs, whilst ensuring collaboration amongst State, local communities, private sector and society (Eggenberger and Partidário 2000). In this scenario, also the integrated management of water and environmental resources plays an outstanding role, especially in terms of political and administrative boundary spanning, so as to overcome conflicting institutional competences (Mitchell 2005; Kidd and Shaw 2007; Woltjer and Niels 2007). In fact, the Integrated Water Resources Management paradigm, European water policies and directives and, more specifically, their transposition and implementation onto each national context, have been conducive to a progressive reformulation of relations between geographical, political and administrative boundaries, directly exerting an influence tending to bridge the policy gap between integrated water resource management and territorial planning.

The broad range of actions and the role taken by RC regarding the implementation of water policies at the local level, veer towards a careful reflection of the horizontal and vertical relationships that these voluntary contractual agreements are able to establish with the instruments of urban and territorial planning and those of river basin planning. In this perspective, RC present great potential since they can take as much a vertical extension, involving multiple administrative levels of territorial government, as a horizontal one, gathering around a single action plan numerous institutional stakeholders. Therefore, in more general terms RC act as important catalysts for networks of concerted efforts between the different institutional levels involved in the governance of a fluvial territory. From this point of view, RC as negotiated programming tools, should reach a greater degree of interrelation with territorial planning and river basin requalification processes, since they often have a decisive role in matters of resolution of conflicting local interests (Magnaghi 2008, 2011; Brun 2014).

As far as planning is concerned, it is acknowledged that RC are devoid of binding power from an urban-planning and territorial standpoint, even though actually dealing with the issues of environmental protection and sustainable development. In fact, on the one hand, the current legal profile of RC and, on the other hand, the routes taken for their development, often independently of the sanctioned administrative procedures for territorial planning tools, are two underlying reasons behind the misalignment with the territorial programs and plans adopted at the higher-level government. These aspects also explain the scarcity of technical and scientific literature on the subject, as well as the dearth of references to the instruments of urban and territorial planning that characterize these contracts, both in contents and documentation.

In France, aspects of integration between RC and urban and territorial planning have assumed a particular relevance, stimulating reflection conducive to more advanced legislation geared to the appropriate realignment between the respective development stages and concrete actions.

In this direction, the enactment of Law No. 338 of April 21, 2004, transposing Directive 2000/60/EC, dictates that urban instruments on any scale must be compatible with the orientations of integrated management of water resources as defined by water management plans at the river basin and sub-basin levels, respectively via *Schéma Directeur d'Aménagement et deGestion des Eaux (SDAGE)* and *Schéma d'Aménagement et de Gestion des Eaux (SAGE)*. This legislation underscores, in this regard, the requirement to weigh the impacts induced by planning tools on the system of surface waters and ground waters (Gouritin 2012).

The case of France is therefore paradigmatic for analyzing the interrelationships between different areas of government of the territory. In fact, a first look at this context reveals how the problem of the interaction between territorial planning policies and water policies is not only that one of geographical and associated physiographic perimeters, but also that one of the actual degree of coherence reached between programming and managerial practices by each planning tool.

It is therefore a matter of exploring the limits that the areas of water management and urban and territorial planning impose on each other, highlighting the real obstacles to full integration between the two practices, even though a more thorough understanding of the objectives and purposes of either (Hellier et al. 2009). This must translate into the activation of further, even interdisciplinary, synergies between the actors of water management and those of urban and territorial planning, so as to more fully explore the opportunity of modeling newer planning tools onto territorial scales more consistent with those of the river basin.

At any rate, a lack of effective coordination between planning practices and management of water resources currently represents a limit to the legal and functional extension of RC towards truly integrated planning of hydrographic territories, as is the case, for example, in Italy.

As a matter of fact, the most natural pathway for integration between urban, territorial and water resource planning cannot clearly discount an analysis of the relations linking urban and territorial context to hydrographic units. In this regard, it is noteworthy that basin plans represent higher-level programming and management tools, with respect to urban and territorial instruments, and it is, thus, within the river basin planning that RC, as such constituting its executive acts, can find a space to mutually realign themselves with the others planning tools.

Again, the French context, in this regard, provides a concrete example of this natural pathway of integration, which emerges by directly observing the relations between urban and territorial planning, and river basin plans.

In the case of the Belgian region of Wallonia the theme of integration between water resource management tools and plans for the government of the territory, is characterized by a greater number of interconnections between RC and both urban and territorial planning, besides of course with river basin management plans. Specifically, some correlations exist with the *Code Wallon de l'Aménagement du Territoire, de l'Urbanisme et du Patrimoine* and the ensuing programming tools, including the *Plan Communal d'environnement pour le Développement Durable (PCEDD)*, the *Programme Communal de Développement Rural (PCDR)*, and the *Plan Communal de Développement de la Nature (PCDN)* (Tricot et al. 2001).

Generally speaking, the very characteristics of RC (contractual approach, partaking on a voluntary basis, vertical and horizontal extension of the institutional context, multi-sectoral scope of actions, participatory approach, local implementation of European and national water policies) highlight its broad capacity in terms of flexibility and applicability to different territorial and socio-economic contexts, and politico-administrative settings. By virtue of this potential, they may also take on an important connecting function between river basin plans and processes of urban and territorial planning, also in the prospects of an evolution of the latter and of a greater coherence between territorial scales on which both planning practices are based.

Although still characterized by a number of legal limits, as well as time-wise, in relation to the purported objectives, RC represent valid tools for the implementation of policies of integrated management of waters on a local scale and furthermore for the innovation in territorial governance.

References

Allain S (2004) Contrats de rivière et agriculture: quel pouvoir incitatif et quelle efficacité environnementale. Revue européenne de droit de l'environnement 4:401–413

Aubin D, Varone F (2001) La gestion de l'eau en Belgique. Analyse historique des régimes institutionnels (1804–2001). Courrier hebdomadaire du CRISP 26–27(1731–1732):5–75

Bastiani M (2011) Contratti di Fiume. Pianificazione strategica e partecipata dei bacini idrografici. Approcci, esperienze, casi studio. Dario Flaccovio Editore, Palermo

Billet P (2008) Les approches volontaires et les principes du droit de l'environnement. In: Hervé-Fournereau H (ed) Les approches volontaires et le droit de l'environnement. Presses universitaires de Rennes, Rennes, pp 97–109

Biswas AK (2004a) Integrated water resources management: a reassessment. Water Int 29(2): 248–256

Biswas AK (2004b) From Mar del Plata to Kyoto: a review of global water policy dialogues. Global Environmental Change Part A(14): 81–88

Bobbio L, Saroglia P (2008) Lungo il fiume delle politiche contrattualizzate. Dall'esperienza francese al caso piemontese. Laboratorio di Politiche-Corep, Torino, pp 1–28

Brun A (2010a) Les contrats de rivière en France: un outil de gestion concertée de la ressource en local. In: Schneier-Madanes G (ed) L'eau mondialisée. La gouvernance en question, Éditions La Découverte, Paris, pp 305–321

Brun A (2010b) Les contrats de rivière en France: enjeux, acteurs et territoires. Les Cahiers de droit 51(3–4):679–704

Brun A (2014) France's water policy: the interest and limits of river contracts. In: Schneier-Madanes G (ed) Globalized water: a question of Governance. Springer, Dordrecht, pp 139–147

Brun A, Lasserre F (2006) Politiques de l'eau. Grands principes et réalités locales. Presses de l'Université du Québec, Québec

Brun A, Marette S (2003) Le bilan d'un contrat de rivière: le cas de la Reyssouze. Économie rurale 275:30–50

Campos E, Sánchez L, Soriano L (2011) Análisis socio ambiental de la cuenca del Matarraña y aspectos técnicos para la elaboración de un contrato de río. VII Congreso Ibérico sobre Gestión y Planificación del Agua "Ríos Ibéricos +10. Mirando al futuro tras 10 años de DMA", 16/19 de febrero de 2011, Talavera de la Reina

Carter JG (2007) Spatial planning, water and the water framework directive: insights from theory and practice. The Geographical Journal 173(4):330–342

Delmas MA, Terlaak A (2001) Voluntary agreements for the environment: institutional constraints and potential for innovation. In: Orts EW, Deketelaere K (eds) Environmental Contracts. Comparative approaches to regulatory innovation in the United States and Europe. Kluwer Law International, The Hague, London, Boston, pp 349–367

Dervieux A (2005) La difficile gestion globale de l'eau en Camargue (France): le Contrat de delta, VertigO - la revue électronique en sciences de l'environnement [En ligne] 6(3)

Eggenberger M, Partidário MR (2000) Development of a framework to assist the integration of environmental, social and economic issues in spatial planning. Impact Assessment and Project Appraisal 18(3):201–207

Enserink B, Patel M, Krantz N, Maestu J (2007) Cultural factors as co-determinants of participation in River Basin Management. Ecol Soc 12(2)

Faure M (2001) Environmental contracts: a flemish law and economics perspective. In: Orts EW, Deketelaere K (eds) Environmental Contracts. Comparative Approaches to Regulatory Innovation in the United States and Europe. Kluwer Law International, The Hague, London, Boston, pp 167–178

Gailliard N, Barreteau O, Richard-Ferroudji A (2014) Participatory policy making in practice: simulating boundary work in Water Governance. In: Kaminski B, Koloch G (eds) Advances in Social Simulation, Advances in Intelligent Systems and Computing 229. Springer, Berlin, pp 333–342

Gouritin A (2012) Water management and protection in France. In: Alberton M, Palermo F (eds) Environmental Protection in multi-layered systems: comparative lessons from the water sector. Martinus Nijhoff Publishers, pp 433–462

Giordano MA, Wolf AT (2003) Sharing waters: post-rio international water management. Nat Res Forum 27:163–171

Hervé-Fournereau H (2008) Les approches volontaires et le droit de l'environnement. Presses universitaires de Rennes, Rennes

Hellier E, Carre C, Dupont N, Laurent F, Vaucelle S (2009) La France, la ressource en eau – usages, fonctions et enjeux territoriaux. Armand Colin, Paris

Hooper B (2005) Integrated River Basin Governance, learning from international experience. IWA Publishing, London, Seattle

Jeffrey P, Gearey M (2006) Integrated water resources management: lost on the road from ambition to realization? Water Sci Technol 53(1):1–8

Kaczmarek B (2003) Politiques communautaires de gestion par bassin. In: Drobenko B (ed) Vers une stratégie de gestion durable des fleuves. P U de Limoges, Limoges Cedex, pp 113–125

Kaika M (2003) The water framework directive: a new directive for a changing social, political and economic European framework. Eur Plan Stud 11(3):299–316
Kaika M, Page B (2003a) The EU water framework directive: part 1. European policy-making and the changing topography of lobbying. European Environment 13:314–327
Kaika M, Page B (2003b) The EU water framework Directive: part 2. European policy-making and policy shifting choreography of governance. European Environment 13:328–343
Kidd S, Shaw D (2007) Integrated water resource management and institutional integration: realising the potential of spatial planning in England. Geogr J 173(4):312–329
Magnaghi A (2008) I contratti di fiume: una lunga marcia verso nuove forme integrate di pianificazione territoriale. Notiziario dell'Archivio Osvaldo Piacentini 1, Reggio Emilia, pp 89–98
Magnaghi A (2011) Contratti di fiume e pianificazione: uno strumento innovativo per il governo del territorio. In: Bastiani M (ed) Contratti di Fiume. Pianificazione strategica e partecipata dei bacini idrografici. Approcci, esperienze, casi studio. Dario Flaccovio Editore, Palermo, pp 31–44
Mitchell B (2005) Integrated water resource management, institutional arrangements, and land-use planning. Environ Plann A 37:1335–1352
Molle F (2006) Planning and managing water resources at the River-Basin level: emergence and evolution of a concept. (IWMI Comprehensive Assessment Research Report 16) Colombo, Sri Lanka, International Water Management Institute
Monge C (2015) Contratos de río. Presented at the 2015 UN-Water Annual International Zaragoza Conference 15–17 Jan 2015
Monge C, Presa C (2011) Condiciones necesarias para la puesta en marcha de un contrato de río en el Matarraña. VII Congreso Ibérico sobre Gestión y Planificación del Agua "Ríos Ibéricos +10. Mirando al futuro tras 10 años de DMA", 16/19 de febrero de 2011, Talavera de la Reina
Moss T (2004) The governance of land use in river basins: prospects for overcoming problems of institutional interplay with the EU water framework directive. Land Use Policy 21(1):85–94
OECD (2007) OECD Environmental performance reviews Switzerland
OECD (2010) OECD Environmental performance reviews Luxembourg
Orts EW, Deketelaere K (2001) Introduction: environmental contracts and regulatory innovation. In: Orts EW, Deketelaere K (eds) Environmental contracts. Comparative approaches to regulatory innovation in the United States and Europe. Kluwer Law International, The Hague, London, Boston, pp 1–35
Rahaman MM, Varis O (2005) Integrated water resources management: evolution, prospects and future challenges. Sustainability: Sci Pract Policy 1(1):15–21
Rahaman MM, Varis O, Kajander T (2004) EU water framework directive vs. integrated water resources management: the seven mismatches. Water Res Dev 20(4):565–575
Rosillon F, Lebeau J (2010) La GIRE décryptée. Eléments pour un renforcement de la GIRE en Haïti et dans les Pays en développement. PROTOS, Université de Liège, Département des Sciences et Gestion de l'Environnement, Arlon (Belgique)
Rosillon F, Vander Borght P (2001) Mobilisation au fil de l'eau: dix années d'expérience de CdR en Wallonie. Annales de Mines, pp 39–54
Sancy M (2008) Quelques réflexions sur l'utilisation des approches volontaires et le droit International de l'environnement. In: Hervé-Fournereau N (ed) Les approches volontaires et le droit de l'environnement. Presses Universitaires de Rennes, Rennes
Shen D, Varis O (2000) World water vision: balancing thoughts after The Hague. Ambio 29 (8):523–552
Solanes M, Gonzales-Villareal F (1999) The Dublin principles for water as reflected in a comparative assessment of institutional and legal arrangements for integrated water resources management. TEC background paper No. 3, Global Water Partnership, Stockholm, Sweden
Teodosiu C, Barjoveanu G, Teleman D (2003) Sustainable water resources management. 1. River Basin Management and the EC water framework directive. Environ Eng Manage J 2(4):377–394

Tippet J, Searl B, Pahl-Wostl C, Rees Y (2005) Social learning in public participation in river basin management. Environ Sci Policy 8(3):287–299
Tricot B, Lejeune-Delforge A, Nuttens B, Rossillon F, Vander Borght P (2001) Guide méthodologique relatif au contrat de rivière. Imprimerie de la Région Wallonne
UNEP (United Nations Environment Programme) (1972) Declaration of the United Nations Conference on the Human Environment, Stockholm 1972
Woltjer J, Niels A (2007) Integrating Water management and spatial planning strategies based on the Dutch experience. J Am Plann Assoc 73(2):211–222

Chapter 3
Comparative Analysis Between France and Italy

Abstract In this chapter, the role of river contracts in European and national water policies has been analyzed, by observing the two paradigmatic contexts of France and Italy. The comparative analysis undertaken both with respect to the experiences of water management policies and river contracts, allows to more deeply assess the governance processes of water resources and understand their political, legal and socio economic scopes. In addition, the comparison of these two national contexts has highlighted the opportunities for actual integration between practices of urban and territorial planning, and water management policies. The analysis has been specifically focused on the respective normative frameworks, water management tools, actors involved, and varying territorial and administrative structures.

3.1 Introduction

In order to analyze the role of river contracts (RC) in European and national water policies and their contribution to innovation as regards the processes of urban and territorial planning, the two paradigmatic contexts of France and Italy were chosen as the focus of the research. The decision to investigate the two national and regional scenarios of these two countries, via a comparative research approach, is based on a number of core considerations.

In the protracted process leading to the adoption of Directive 2000/60/EC, European legislation followed the example of France, long considered a country of reference in matters of water integrated management policies and practices (Scarwell and Laganier 2004; Ghiotti 2006; Pezon 2006). In fact, in France, the normative, administrative and procedural traditions for water resource and territorial management date back to the mid-1960s, being characterized right from the start by a clear recognition of the hydrographic basin as the optimal unit of planning and management (Ghiotti 2006). The rich and diversified French scenario presents numerous operational tools at the river basin or sub-basin scale, such as the *contrat de rivière* (*CdR*) introduced at the beginning of the 1980s, as a result of thirty years

M.L. Scaduto, *River Contracts and Integrated Water Management in Europe*,
UNIPA Springer Series, DOI 10.1007/978-3-319-42628-0_3

applied experience in the field of negotiated and participatory processes of water resources management (Dervieux 2005).

In comparison, the overall picture in Italy appears much less evolved and as yet to be defined both in terms of its national policy and of its implementation tools on the local scale (Bianco and Pineschi 2011). Its underdeveloped evolutionary stage represents a matter of particular interest towards the Italian context, especially if combined with the relatively recent adoption of the RC paradigm, adopted in Italy since the early 2000s.

In this sense, the comparative analysis of the experiences of water management policies and RC in France and Italy is able to contribute, on the one hand, to a broader and more in-depth assessment of the governance processes of water resources, based on concerted, negotiated and participatory instruments, as well as to a greater understanding of their political, legal and socio-economic scopes, on the other. In addition, comparison of the two scenarios can more clearly disclose the opportunities for actual integration between practices of urban and territorial planning and water management policies.

The research was structured on two levels of analysis of the French and Italian contexts, a more general one concerning water management policies and a more specific one regarding RC.

As to the former level of investigation, a comparison of the corresponding normative references, management and applications tools and actors involved in the policies of water protection and management has been undertaken, always taking into account the varying territorial and administrative structures and the diversity in outcomes elicited by the implementation of the European Water Framework Directive (WFD).

With reference to the latter level of comparison, four specific aspects of the application of RC were analyzed in parallel: (I) legal and regulatory references, (II) contents and procedures of the contractual agreements, (III) actors involved, (IV) implementation experiences concluded or underway.

Where applicable, the dual-level comparative analysis was integrated with references to other national contexts, so as to take a wider cognizance of the dissemination and scope of RC in the European scenario.

3.2 Comparison of Water Management Policies

The comparative analysis provides for first considering the lengthy evolutionary process of water management policies in France and then shifting focus to the peculiarities of the Italian context.

In France, the long-standing tradition in the field of integrated water management descends from laws of the second half of the nineteenth century (Bravard 2002), evolving through the two fundamental *Lois sur l'eau* of 1964 and 1992, and being completed in 2006 with the *Loi sur l'eau et les Milieux Aquatiques* (*LEMA*) (Brun 2014). Until the mid-1960s, the management of water resources continued to be handled according to a sectoral logic that was less integrated with environmental issues (Scarwell and Laganier 2004), whereas in recent years an approach more focused on the hydrographic units has been achieved (Laganier et al. 2009). The latter units were defined by Law No. 1245 of December 16, 1964 "*Relative au régime et à la répartition des eaux et à la lutte contre leur pollution*" that acknowledged the importance of the decentralized water resources management at the river basin scale and the interdependencies between the different uses, and the subdivided the entire national territory into six large metropolitan river basins.

The management of each river basin is entrusted to a *comité de bassin*, which represents a veritable *parlement régional de l'eau* with consultative and deliberative functions, and an *agence financière de bassin*, a body whose functions are executive as well as to promote cooperation between territorial actors (Larrue 2002; Barraqué 2003; Nicolazo and Redaud 2007).

The *comités de bassin* are composed in equal parts of representatives and experts of local authorities and central administrative authorities of the State as well as of the various categories of water users, under article 13 of the same law of 1964. In addition, these committees cooperate with the *commissions géographiques* present in major hydrographic basins, representing venues for consultation and discussion between local stakeholders (Scarwell 2007).

The *agences financière de bassin* have both technical and financial authority, and their institutional mission is to effectuate the policies set by the *comités de basin* to improve quality and quantity of water resources, via long-term programs.

With reference to the economic aspects regarding the management of water resources and the corresponding hydrogeological and environmental risks, the first *Loi sur l'eau* of 1964 anticipated subsequent regulation concerning water pollution by introducing the financial mechanism based on *payeur-pollueur*, i.e. the principle whereby resource consumption, environmental impact and the corresponding tax liability of users are to be directly proportionate (Scarwell 2007).

The same law, taken as a reference by many other European countries, has been of considerable importance in the evolution of the French *politique de l'eau* for two main reasons: (I) its marked orientation towards the river basin scale regards the integrated management of surface and underground waters; (II) the requirement that all stakeholders participate in decision-making processes and implementation of programs (Larrue 2002; Massardier 2009). With regard to participatory processes, the first *Loi sur l'eau* ushered in a new course leading to the optimization of decision making processes that, beginning in the late 1970s on the initiative of the *Conseiles generaux*, found a further element of innovation in the institution of the *Établissement Publics Territoriaux*. The latter, in the capacity of *syndicats mixtes* composed of local authorities, even answering to diverse *départments*, have themselves become executors of national political orientations, through numerous programs of actions conducted at river basin and sub-basin scales.

Another turning point was marked by the second *Loi sur l'eau* of January 3, 1992 which, in direct continuity with the law of 1964, defined water «*patrimoine commun de la Nation*», thus *res communis* (Barraqué 1994, 374), thereby sanctioning for the first time the transversal nature of water policy at the river basin scale

(Laganier et al. 2009). In this sense, this law marked the transition from the concept of water as a mere resource to an eco-systemic vision of water values (Puech and Boisson 1995; Ghiotti 2001; Larrue 2002).

From the operational standpoint, the *Loi sur l'eau* of 1992 introduced two new water management schemes, the *Schéma Directeur d'Aménagement et de Gestion des Eaux* (*SDAGE*) and the *Schéma Aménagement et de Gestion des Eaux* (*SAGE*). The former, represents a mandatory plan for each of the six large basins and, as such, superordinate instrument with respect to other urban and territorial planning tools, in parallel the *SAGE* is a non-compulsory scheme for the management of individual sub-basins.

For the practical application of *SDAGE* and *SAGE*, the second *Loi sur l'eau* also sanctioned the role of negotiated and participatory procedures in the implementation of integrated water planning, among which RC were duly comprised (Scarwell 2007; Laganier et al. 2009; Massardier 2009; Brun 2014). According to some authors, the successful outcomes obtained by way of *CdR* were indeed the very conditions that encouraged legislators to integrate the *SAGE* in the second *Loi sur l'eau* of 1992 as a new model of territorial negotiation precisely inspired by the same RC (Lascoumes and Le Bourhis 1998).

Development and updating of the *SDAGE* are the remit of the *comités de bassin*, which define its general orientation regarding water protection and management. At the same time, the *Agences de l'eau* and the *Directions Régionales de l'environnement, de l 'Aménagement et du Logement* (*DREAL*) co-ordinate and oversee the implementation of the established orientations, in close synergy with local actors of water management, representatives of professional communities and water users.

While *SDAGE* outline general policy lines regards the planning and management of water resources, *SAGE* aim at resolving existing conflict arising from the contrasting interests within river basins or sub-basins contexts (Larrue 2002; Flajolet 2006). In fact, the development, implementation and coordination of *SAGE* policies encompass the strong political will of the local actors partaking in the *SAGE* control room, by way of their participation in the *Commissionnes Locaux de l'Eau* (*CLE*) (Briola 2004).

The latter are the territorial commissions wherein the concerted decision-making processes unfold. The *CLE* is made up by representatives of local authorities (50 %), representatives of users and associations (25 %), and by representatives of state organizations and its *établissements publics* (25 %), and unlike the *comités de bassin*, are characterised by a broader representation on behalf of local stakeholders (Larrue 2002).

The active involvement of local actors specifically assumes a marked significance during the *SAGE* development stages (Larrue 2002; Le Bourhis 2003), during which the status of hydrographic territories must be opportunely considered and the appropriate planning scenario accordingly defined, even by means of alternative proposals and according to a variety of sustainability indicators. The active participation of all stakeholders represented in the *CLE* is therefore crucial to devising a strategy capable of achieving the *SAGE* objectives, as well as for the follow up and monitoring of the efforts undertaken.

The regulatory framework outlined by the second *Loi sur l'eau* of 1992 was recently updated and amended by Law No. 338 of April 21, 2004, which transposed the European Water Framework Directive into France, establishing 9 national hydrographic district (plus 5 for over-sea provinces) (Fig. 3.1). This law specifically identifies two authorities to whom are referred the main functions in the field of integrated water resources management at the river basin scale: (I) the *comité de bassin* which, in continuity with earlier legislation, is entrusted with impact analyses, programming and control measures, ensuring that policies and actions are in conformity with the river basin district management plan, elaborated by the committee itself; (II) the *préfet coordonnateur de bassin* who approves and enforces the river basin district plan. Furthermore, in line with the WFD precepts on integration between water management and territorial planning, the same law of 2004 calls for

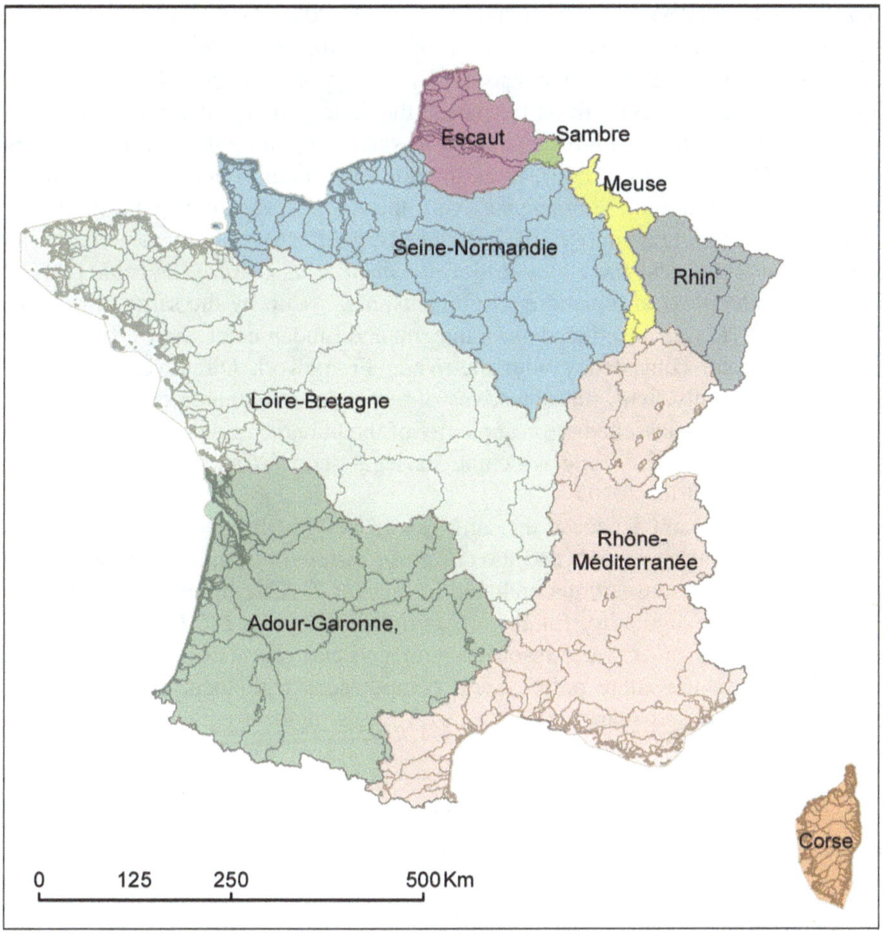

Fig. 3.1 The 9 national hydrographic districts of France

the adaptation of plans such as *Schémas de Cohérence Territoriale* (*SCoT*), *Plans Locaux d'Urbanisme* (*PLU*) and *Cartes Communales*, to the water planning policies defined within the *SDAGE* and *SAGE*.

More recently, the French regulatory framework concerning water policies was further updated by the *Loi sur l'eau et les Milieux Aquatiques* (*LEMA*) of December 30, 2006 (Ghiotti 2010). The *LEMA* reconfirms the importance of decentralized management and governance of water resources at the hydrographic basin scale, identifying the technical measures and methodologies needed to achieve the objectives of the European Water Framework Directive. In addition, this law also introduced some important innovations in relation to: (I) the composition of the *comité de bassin*, with a larger proportion of elected state and local representatives as well as of water users, (II) the role of the *agences de l'eau* and the relative system of funding, (III) the powers vested in the *comités de bassin*, in matters of water planning, and (IV) the creation of the *Office National de l'eau et des Milieux Aquatiques* (*ONEMA*) (Drobenko 2007; Clarimont 2009; Ghiotti 2010). The latter is a national public agency, overseen by the *Ministère de l'Ecologie et du Développement Durable*, and consists of nine regional offices. The creation of *ONEMA* bespeaks a clear will on the part of the State to more effectively coordinate water programs and initiatives conducted at various institutional and management levels (Caudal and Janin 2007), even though some authors have highlighted the potential for overlapping competences with the *Agences de l'eau* (Ghiotti 2010).

The most important innovation introduced by *LEMA* is about the relationship between the *SAGE* and the *Plans d'Aménagement et de Gestion Durable* (*PAGD*) regarding waters and of aquatic environments, and set up by the same law. In this sense, the *PAGD* was introduced as an integral and fundamental part of the *SAGE* to ensure its greater consistency with the Water Framework Directive, the *SDAGE* and, more generally, with the objectives of territorial sustainable development. In addition, the *LEMA* lessened the complexity of the bureaucratic procedures required to elaborate the *SAGE*, while widening its legal scope (Caudal and Janin 2007; Clarimont 2009; Brun 2014).

Another significant aspect of the same law of 2006 is the relevance attributed to participation of private actors and the citizenry, fostered through the dissemination of objectives, rules and actions included in the *SAGE* (Clarimont 2009). From the perspective of the regulatory framework as redefined by the *LEMA*, therefore, new forms of governance of water resources emerged and revealed themselves crucial factors to harmonize State policies and competences with local territorial development programs and plans, and thus, with needs of local agencies and communities (Scarwell 2007; Ghiotti 2010).

In this vision, the *SAGE* and the *CdR* simultaneously represent a procedural route, as operational measures and inclusive participatory processes, particularly versatile and useful for the local implementation of European Community (EC) and national water policies. On the other hand, these two kinds of operational management instruments, also due to the simplification effects on *SAGE* introduced by the *LEMA*, have become increasingly complementary, albeit maintaining some distinguishing features that set them apart in terms of their legal nature, time requirements for their development, duration of action plans and potential discernible advantages for the territory. The progressive alignment of *SAGE* with the *CdR*, is also an incentive for the adoption of the former even in those territorial contexts in which RC have been favored due to their lower bureaucratic, technical and legal complexity, and therefore greater versatility in comparison to the various local politico-institutional, administrative and local socio-economic contexts (Allain 2001; Clarimont 2009). To this regard, the Table 3.1 offers a synoptic comparison highlighting the salient features of *SAGE* and *CdR*, from which can be inferred the different degree of complexity that characterizes these two types of operational measures.

In the most recent phases of its evolutionary pathway, the picture of French national water policies has established some fundamental interconnections with environmental policies. From this standpoint, the two *Loi Grenelle* of 2009 and 2010 (no. 967 of August 3, 2009 and no. 788 of July 12, 2010), have traced a significant convergence between these two policy sectors, thanks to the introduction

Table 3.1 Synoptic table of *SAGE* and *Contrat de rivière* elements

	SAGE	Contrat de riviére
Type	Technical instrument for planning and management of water resources with a 10–15 years validity, defining a system of actions concerted within the *Commission Locale de l'Eau*	Technical and financial instrument for implementing a program of actions to safeguard and manage a river, its resources and territory, with a 5–7 years validity, elaborated and monitored by *syndicats de rivière* and *comités de rivière*
Legal nature and scope	Technical sectoral plan whose legal scope is superordinate to urban and territorial planning instruments	Voluntary contractual agreement between local stakeholders assuming different roles of sponsor, executors and financial supporters
Subject in charge of decision-making, consultation and monitoring	*Commission Locale de l'Eau* composed of 50 % local political representatives, 25 % water users, 25 % other administrations representatives	*Comité de rivière* with a variable composition of, representatives of the involved cosignatories
Role of the Prefect du département	Ratifies the definition of the SAGE perimeter, the composition of the *CLE* and the content of action program	Defines the composition of the *comité de rivière*, ratifies and underwrites RC on behalf of the State
Territorial area of reference	Hydrographic sub-basin	Hydrographic sub-basin or its sub-units
Time requirements to design action plan	At least 4–5 years	Averaging 3 years
Relations with other water management tools	May be implements in conjunction with one or more *CdR*	May facilitate the implementation of a *SAGE*

of the so-called *trame verte et bleue* and also by fostering a greater degree of interaction between land-use planning and water management. Specifically, the theme of the *trame verte et bleue* presents great potential for direct integration with tools for planning and management of water resources, since the *blue* element of this ecological weave is represented by the hydrographic network and the natural habitats closely related to it, such as wetlands (Brun 2014).

At local level, in the framework of these new interactions between environmental protection policies and integrated water resource management, the *SAGE* and the *CdR* can be subjected to a well-timed update in terms of their objectives, contents and procedures, thus broadening their legal scope and applicability to many other different territorial contexts.

In Italy, in line with the trends in France and Europe the evolution of water management policies was framed around a number of legislative milestones and dragged along by the rather slow and complex tide of reform regarding institutional reorganization and decentralization of State powers to local authorities.

An analysis of the legislative cornerstones put in place from 1933 to 2006, allows one to retrace the path already outlined in the literature (Goria and Lugaresi 2004), and to update it opportunely in the light of the transposition of the WFD into the code on *Norme in materia ambientale*, i.e. environmental rules and regulations (Legislative Decree No. 152 of April 3, 2006).

The historical picture of water management policies can be subdivided into three main periods, each characterized by predominant legal and procedural features.

The beginning of the first period coincides with the issuing of the *Water Code* of 1933 (Royal Decree No. 1775/1933) and lasts until the coming into effect of Law No. 319 of 1976 on the safeguard of water resources from pollution. Protection of the population and territories from hydrogeological risks represented, over this long stretch, the predominant concern of Italian water policy. The crucial point of this national policy was paradoxically marked during the post-World War II period, by catastrophic floods that in the 1950s and 1960s affected extensive rural districts (the wide flood of 1951 in *Polesine* region), as well as important urban and historical-cultural contexts (the great flood of 1966 in Florence).

The onset of the subsequent historical phase was occasioned by these devastating natural disasters and by the protracted and complex political, scientific and technical debates that ensued. These debates led to two important turning points in the policies regarding the protection and management water resources, and the safeguard of territories, marked by the aforementioned Law No. 319 of 1976, as well as the Law No. 183 of 1989 on the *Norme per il riassetto organizzativo e funzionale della difesa del suolo* i.e. regulation for the functional and organizational restructuring in matter of soil protection (Di Federico 2003; Ferrucci 2003; Zazzi 2009). Specifically, the importance of the former lies mainly in the introduction of certain sustainability features into water resource management (water waste control, discharge planning, regulation of water uses, resources protection), all the more in a historical juncture marked by the transfer of State competences in water matter to regional administrations, established in the second half of the 1970s (Law No. 616/1977). On the other hand, the scope of Law No. 183 of 1989 was definitely wider, since it integrated into the existing regulatory framework some innovative aspects regarding the protection of the hydrographic network and its water resources, besides the management and regulation of their uses. Indeed, this law sanctioned: (I) the identification of the hydrographic basin as the territorial unit of reference for integrated water planning measures and programming interventions of a broader nature than mere flood and erosion control alone; (II) the establishment of the Italian River Basin Authorities, agencies in which both State and Regions have stakes, characterized by autonomy in terms of competences and financial resources; (III) the introduction of River Basin Management Plans, conceived as land-use sectoral planning instruments and defined as «means to increase knowledge for improved decision making, for regulatory and operational technical purposes, and for planning and programming actions and regulations aimed at preserving, defending and enhancing the land and the proper use of water resources, based on the physical and environmental characteristics of each territory» (Law No. 183/1989, art. 17) (Goria and Lugaresi 2004; Menduni 2007).

From the standpoint of Law No. 183/1989, the hydrographic basin and the River Basin Management Plan were recognized as fundamental planning bases, both for water management and for soil protection, overcoming the approach previously in force based on distinct territorial administrative areas. However, in light of the complexity of the measures required for increasing knowledge, programming and planning, the same law provides for the eventuality that River Basin Management Plans be drafted and approved in form of *piani stralcio* i.e. plans to be specifically implemented for sub-basins or similar functional areas (art. 17, comma 6-7 ter). The so called *piani stralcio* were also introduced to allow the newly-established Basin Authorities to, first and foremost, consolidate their know-how, competences and capacities in terms of institutional and territorial consultation and coordination, which all represent crucial elements for drafting basin-plans (Ferrucci 2003; Ercolini 2006; Filpa 2009).

Within this second historical phase of Italian regulatory framework evolution, two other legislative cornerstones were put in place aimed at consolidating a more integrated management of water resources. The first was law No. 36 of January 5, 1994 *Disposizioni in materia di risorse idriche* i.e. provisions concerning water resources, followed by Legislative Decree No. 152 of May 11, 1999 *Recante disposizioni sulla tutela delle acque dall'inquinamento e recepimento della direttiva 91/271/CEE concernente il trattamento delle acque reflue urbane e della direttiva 91/676/CEE relativa alla protezione delle acque dall'inquinamento provocato dai nitrati provenienti da fonti agricole* i.e. provisions concerning the protection of waters against pollution and the transposition of Council Directive 91/271/EC on urban waste water treatment and Council Directive 91/676/EC concerning the Protection of Waters against Pollution caused by Nitrates from Agricultural Sources. Specifically, the Law No. 36/1994 introduced two new elements in terms of streamlining the management of water services, also in consideration of the inherent administrative and territorial complexity of hydrographic basins. In this regard, the law established the *Ambiti Territoriali Ottimali* (*ATO*) and *Autorità d'Ambito Territoriale Ottimale* (AATO) in order to minimize the host of individuals involved in the management of water services, overcome the fragmentation of agencies and management processes, and create economies of scale for the benefit of local agencies, but especially of water users. Under this legislation, the Regional administrations were attributed the vested powers and the burden of identifying the territories comprising the ATO, establishing the competent Authorities, levying water tariffs, as well as identifying the most appropriate forms of institutional cooperation (consortia, metropolitan areas, etc.).

Subsequently the Regions were likewise attributed the responsibilities regarding the safeguard of water resources and the task of compiling *Regional Water Protection Plans* (*PRTA*), as established by Legislative Decree No. 152/1999. In particular, the PRTA is a sectoral plan of the basin plan, aimed at reducing pollution, restoring water bodies to a healthy state and improving the quality status of waters, particularly drinking water (Goria and Lugaresi 2004). Its elaboration is a concerted effort on the part of national and interregional Basin Authorities, called on to define the objectives and priorities of the protection plans on the basin scale, and the Regions that prepare and adopt the PRTA, and then transmit it to the competent Basin Authorities for approval upon verification of its compliance with the objectives and actions envisaged.

Following the amendments introduced by the Legislative Decree No. 152/1999, the evolution of the regulatory framework concerning the protection and management of water resources has outlined new prospects in the structure of the relationship between river basin and urban and regional planning practices. Specifically, on the one hand basin plans retain their role as superordinate strategic plans, setting the objectives, priorities, implementation timetables and financial flows; on the other hand, PRTAs represent a framework that takes into account the physical, environmental and anthropic contexts of each hydrographic body, as well as detailed action programs, while constituting useful technical and operational tools for identifying causes of many forms of environmental and water pollution. In this sense, PRTAs contribute to more accurately delineate basin plans and, in particular, to the safeguard of water quality, also through obligations and limitations on behalf of environmental protection (Ferrucci 2003).

The launch of the latest phase of the evolution of the Italian regulatory framework in the field of integrated water resource management, coincides with the enactment of Legislative Decree No. 152/2006 *Norme in materia ambientale*, i.e. environmental rules and regulations. This decree aggregated all the existing legislation on the protection, planning and management of waters while introducing some innovations, first and foremost the institution of eight *Hydrographic Districts* (Fig. 3.2), which include the prior hydrographic areas on the national, regional and interregional territorial scale, thereby transposing in this regard the dictates of WFD. Moreover, the same decree likewise foresaw an equal number of *District Authorities*, to be entrusted with the task of drawing up the District Management Plans, i.e. the so-called *Piani di Gestione del Distretto idrografico* (PGD). The latter represent «the cognitive, regulatory and technical-operational tools, by means of which actions and regulations are planned and scheduled for purposes of preservation, protection and exploitation of soil and for the proper use of water, based on the physical and environmental characteristics of the area concerned» (Legislative Decree No. 152/2006, article 65, paragraph 1), even though, according to some authors, there remain several critical elements relative to its practical application (Menduni 2007; Bonami and Brugioni 2011).

With respect to regulatory updates introduced by Legislative Decree No. 152/2006, the PRTAs continue to represent the implementation tools of the basin plans, renamed district management plans in virtue of the same decree. Consequently, on the scale of each hydrographic district PRTAs represent the *trait d'union* between the strategies defined by each district management plans, and the

Fig. 3.2 The 8 national hydrographic districts of Italy

set of measures envisaged by other water planning instruments at the regional and local levels. In this sense, PRTAs assume the role of the so called *piani stralcio* and reveal themselves necessary for the implementation of the hydrographic district plans.

Some obvious limits of the application of Legislative Decree No. 152/2006 show that the Italian Environmental Code, far from leading to an organic reorganization of the legislation on integrated water resource management and environmental policy, has simply conflated pre-existing parts of the laws it abrogated. For example, certain aspects of the functional and hierarchical inter-relationships between different plans, subjects and institutional levels remain still unsolved and rather confused, and at the same time some of the institutional competences in water management tend to overlap in many cases, often with arising conflicts (Urbani 2007; Bianco and Pineschi 2011). Moreover, this scenario is further aggravated by the lack of all District Authority establishment that, despite the provisions of the legislative decree No. 152/2006, have not yet been duly enacted by a specific law, thereby leaving the task of drawing up the district basin management plans to the extant Basin Authorities or, in the case of Sicily and Sardinia, to the Regional administrations (Zazzi 2009; Bonamini and Brugioni 2011).

Overall, the analysis of the legislative acts that have followed one another, from the Water Code of 1933 to the Legislative Decree 152/2006, shows that the evolution of the Italian regulatory framework concerning the protection and management of water resources has unfolded primarily through regulatory cornerstones, that have been strongly conditioned by the aftermaths of catastrophic events of the second post-World War era and, more recently, by the provisions of the European WFD of 2000. Consequently, the bottom line of this evolutionary pathway is a quite fragmented and fragmented legislative scenario as regards the effective integration between economic and environmental policies, the coordination between different institutional levels and the dialogue between urban-territorial planning practices and those of the integrated water management sector.

3.3 Legislative Frameworks of River Contracts

In France, the regulatory framework of the *CdR* has always been grounded on circulars of the Ministry of Environment that, from 1981 until 2004, have modeled and refined contents and procedures to conform to the evolution of EC and national water resource management policies. These circulars have gradually altered the scope of the *CdR* and have transformed them from procedures that initially targeted the environmental quality standards of small rivers, into sophisticated tools for the integrated management of water resources.

The Circular of February 5, 1981, issued by the *Ministère de l'Environnement et du Cadre de Vie*, actually instituted the *CdR* and the contractual procedures for their implementation, hailing them as more practical alternatives to traditional regulatory procedures, based on specific laws and regulations (Duport 1991). Indeed, in the text of the circular the *CdR* was defined *«un instrument de réalisation des cartes départementales d'objectifs de qualité sur certaines rivière en faisant appel non à la voie réglementaire mais à la vie contractuelle»*.

The main objective of this first rendition of the *CdR* was to promote and achieve the rapid requalification of water resources and the improvement of rivers in an environmentally sound manner, by defining efforts underwritten by all *rivierains*, that is the land owners of riverside properties along a given river or channel. In fact, the prerequisite of *rivierains* active involvement represented a clear encumbrance to the first *CdR* in large river contexts. All the more, the very wording of the ministerial circular underscored the need for broad consensus among local actors, so as to define the tangible requalification objectives to be proposed in these contractual agreements and to implement the relative programs relying on mutual financial commitments, amongst the parties involved.

By and large, an analysis of the circular of 1981 and its ensuing application, highlights a political and managerial vision that in the early 1980s was still based on the geographical entity of single watercourse, and on the territory administrative frame of reference, i.e. the *département*, with an approach still disjoined from the river basin and strongly oriented towards the sectoral biases of water quality, protection of the riverbed and riverbanks, and management of aquatic environments (Brun 2010).

In the light of the positive results of the early *CdR*, on November 12, 1985 the Ministry issued a second circular to promote the constitution of permanent organisms of management for each watercourse. In the premises of this circular, particular significance was given to the exceptional amount of funding allocated by the *Conseils généraux régionaux* for the stipulation of RC, as well as to the wide interest demonstrated by local communities for these negotial and participatory tools, especially due to their capacity to promote processes of requalification of water resources, whilst supporting efforts to foster local development (Duport 1991). With reference to its contents, the circular of 1985 introduced three specific thematic sections (*volets*): (I) quality levels of waters, (II) renaturation, riverbank management and restoration of aquatic environments and river landscapes to a healthy state, and (III) actions aimed at public information about river management forms.

While addressing the *Agences financières de bassin*, the circular of 1985 put particular emphasis on the requirement that *CdR* be consistent with the framework outlined by the preceding circular of 1981 and that actions plans be monitored both while underway as well as after the deadlines for contractual obligations, such as to protract their positive effects. With reference to the latter, for the first times the circular of 1985 put forward the motion of promoting the institution of permanent management structures for each watercourse.

In the early 1990s, the regulatory framework of the *CdR* evolved even with respect to legislative updates in matters of prevention of natural hazards (Brun 2010). At the same time, these developments were determined by the results of a survey conducted in 1986 by the *Comité Interministeriel à la Qualité de la Vie* on the state of neglect of most rivers and on the management of different river environments, which too often proved fragmentary and sectoral. In this perspective, the investigation was focused on the evaluation of the potential and limits of RC, highlighting the need for a stronger and more efficient coordination on the part of territorial advisory groups (Duport 1991).

On May 13, 1991 the Ministry issued a third circular aimed at improving the quality of coastal waters while announcing the State's participation in the *contrats de baie*, the tools set in place to promote best practices regards the management of water quality in polluted coastal areas and, at the same time, to respond to specific demands on the part of local stakeholders. This third circular stated that in these contexts, usually consisting in a river mouth, estuary, or a coastal pond, projects could be managed by a local organism, such as a *syndicat de communes*, a *département*, a *région*, or on the base a shared responsibility amongst some stakeholders.

A prominent aspect of circular of the 1991 was (I) the mandatory compliance of the *contrats de baie* to the *Schémas d'assainissement communaux ou intercommunaux*, that is programs of redevelopment on a municipal or inter-municipal scale, and (II) the obligation of financial consistency of each contract with respect to other negotiated agreements that might already be in force in the same territory, such as *CdR*.

After the enactment of the second *Loi sur l'Eau* of 1992, a noticeable change took place both in procedures and contents of the *CdR*. With the fourth circular on the *CdR* (March 22, 1993) the *Ministère de l'Environnement* launched a new season of these agreements, taking into account planning and management tools instituted by the same law of 1992, and the acknowledgment of the river basin as the preferential territorial unit of management (Brun 2010). This second generation of *CdR* was therefore more oriented towards integrated water resource management, surpassing in this regard the prior almost exclusive focus on improving the quality of waters that had characterized the foregoing experiences.

In line with the gist of the law of 1992, the circular of 1993 expanded the goals and fields of application of the *CdR* and the *contrats de baie*, emphasizing the concept of sustainable development, the role of local communities, the importance of a broad territorial consultation and the need for an integrated water resource management of a more lasting nature. In addition, for the first time the circular of 1993 sanctioned the need to fully integrate the *CdR* into the renewed sectoral planning context of the *Schémas d'Aménagement et Gestion des Eaux* (*SAGE*), along with a broader participation on the part of local communities. As a consequence of this circular, the *SAGE* and the *CdR* became more attuned to each other, with the first representing the regulatory paradigm of water policies at the local level, and the latter advancing to the forefront in the promotion and implementation of the same policies (Brun 2010). Underscoring their complementary disposition, the terms of the circular of 1993 outlined the competences of the *Commissionnes Locaux de l'Eau*, also in terms of the definition, management and monitoring of *CdR*, with particular regard for their consistency with the *SAGE*.

By and large, the circular of 1993 optimized the initialization and realization procedures of *CdR* drawing attention to the required coexistence of three basic prerequisites: (I) an explicit interest on the part of elected public representatives to set a RC in motion, also with the coadjutant support of the *Agence de l'Eau* and the *Conseil General*; (II) the existence of an approved, or at least in the making, *SAGE* to ensure an appropriately laid out *CdR*, both at territorially and concrete objective level; (III) the appropriate selection of the hydrographical units of reference in order to avoid too vast and heterogeneous perimeters from the standpoint of eco-systemic functioning and of institutional competences and management skills required.

Another milestone in the evolution of the regulatory framework concerning the *CdR* is represented by the circular of October 24, 1994, entitled *Plan Décennal de Restauration et d'Entretien des Rivières. Appel aux Contrats de Rivière*, issued by the Ministry of the Environment with the aim of revisiting the procedural requirements of the *CdR*, to better integrate them specifically with the objectives of the *Plan Décennal de Restauration et d'Entretien des Rivières*. Besides resetting the share of public co-financing, this circular updated the structural requirements of the final dossier necessary to activate each *CdR*. The renewed structure comprised thematic sections making specific reference to (I) urban wastewater treatment programs and depollution of effluents of industrial, farming and animal husbandry practices; (II) restoration and renaturation works on waterways, and actions improving the value of aquatic environments and landscapes; (III) works for the protection of inhabited areas against the hydrogeological risks and floods; (IV) integrated management program for fluvial environments.

Overall, the two circulars of 1993 and 1994 marked an important evolutionary shift in the implementation process of *CdR*, assigning a clearer and more decisive role to the representatives of local public authorities and to the institutional consultation among them.

Along the evolutionary trail of the regulatory framework hitherto described, the apogee was marked by the *Circulaire relative aux Contrats de Rivière et de baie* (No. 3 of January 30, 2004) issued upon adoption of European Water framework Directive of 2000. The aim of this fifth circular was to reformulate the procedural aspects of these negotiated and participatory instruments, while improving their coherence with other water planning and management tools, above all the *SAGE*. To this end, the circular of 2004 decentralized the evaluation hurdle for final dossiers of *CdR*, by transferring this assessment task from the *Comité national d'agrément* to each one *Comité de bassin*. In addition, the circular of 2004 attributed to the *Agences de l'Eau* the competences regarding the identification of actions and measures applicable for financing, classifiable under five thematic sections: (I) regulatory enforcement measures to counter pollution in order to improve water quality; (II) restoration and renaturation works on riverbanks and riverbeds, coastlines and areas subject to floods; together with integrated management actions and measures for improving the value of aquatic and marine environments and related landscapes; and also together protection of fish species; (III) works to improve the safety of inhabited areas against of the hydrogeological risks, floods and rising of the sea level; (IV) actions to improve the availability of water and especially drinking water; (V) local promotion and coordination, monitoring and evaluation over the mid and long term actions and programs in the context of *CdR* and *contrats de baie*.

On the whole, based on the analysis of the ministerial circulars hitherto accounted for, it is possible to identify two main evolutionary elements on the slowly unfolding path leading up to the current configuration of the regulatory framework of the *CdR*: (I) with regard to the territorial context of reference, after the enactment of the second *Loi sur l'Eau* of 1992, the river basin and the sub-basin areas were identified as the most appropriate territorial units for the implementation of the *CdR*, also considering the obligatory coherence with other instruments of water resource planning and management; (II) in terms of the political drive underlying the diffusion and application of *CdR*, a transition took place from a first season of experiences mostly promoted by the Ministry of the Environment, to a new generation of *CdR* elicited by a more decisive political initiative on the part of local institutions and stakeholders, as a telltale indicator of the renewed perception of the potential of these instrument, on behalf of local communities in terms of sustainable development.

In Italy the regulatory framework of river contracts appears much more fragmented and devoid of specific State regulatory efforts, also due to the fact that this kind of negotiated and participatory contracts has been introduced more recently into the various territorial contexts.

The sole law in national legislation making any albeit indirect reference to the RC tool can be found in Legislative Decree No. 152 April 3, 2006 *Norme in materia ambientale* in particular in part III dedicated to soil conservation, contrast to the desertification, protection of waters against pollution and water management. Transposing the European Water Framework Directive, this legislative decree sets out that Basin Plans may be drawn up and approved also for sub-basins or particular sectors or functional parts of a river basin (article 65, paragraph 8) and that Basin Authorities promote the active involvement of all interested parties in the drafting, review and updating of River Basin Plans (article 66, paragraph 7). A legislative process is still underway that should lead to the updating of this legal frame and to the inclusion of explicit references to RC intended, according to the current modification proposal, as procedural approaches able to contribute «to the implementation of district planning tools at the scale of the basin and sub-basin areas, as voluntary tools of strategic and negotiated programming that pursue the protection, proper management of water resources and development of fluvial areas, together with hydrological risk control, while contributing to local development of these areas» (Amendment no. 10.56 to bill no. 1541 proposed at the Senate of the Italian Republic). Particularly, a specific legal update is expected for the end of 2015, as a deed linked to the annual national financial law, that is referred to a re-organization of hydrographic districts and relative management instruments included river contracts.

Currently, therefore, RC have yet to be legally acknowledged at the national level and, in addition, they are the object of different renditions in various regional regulatory contexts (Bianchi and Pineschi 2011).

The Region of *Lombardia* was the first Italian regional administration to recognize the RC tool from a regulatory standpoint, incorporating it into the Regional Law No. 26 of December 12, 2003 *Disciplina dei servizi locali di interesse economico generale - Norme in materia di gestione dei rifiuti, di energia, di utilizzo del sottosuolo e di risorse idriche*, i.e. the regulation on local services of general economic interest, waste management, energy, subsoil use and water resources. In particular, this law in article 45, paragraph 9 of title V-*Regulations on water resources* promotes by means of river and lake contracts consultation and policy integration at the basin and sub-basin levels, with the participation of public and private entities, to safeguard and improve the quality of water resources and their related environments, as well as to ensure protection from risk of floods.

The role of RC in water resource management is also attested, albeit indirectly, in Regional Law No. 12 of March 11, 2005 *Legge per il governo del territorio*, i.e. the law for planning and managing the territory. In particular, this law introduced through article 55bis the Strategic Projects for sub-basins, set up by the Regional Council and drawn up through initiatives entailing participatory processes with all local stakeholders. All the more, the *Piano Territoriale Regionale*, i.e. the regional territorial and landscape plan, makes explicit reference to the experiences of RC carried out in *Lombardia*, clearly manifesting the will to systematically rely on the various urban-territorial and sectoral planning tools at the regional level (Bastiani 2011; Clerici et al. 2011).

In the case of the Region of *Piemonte*, the RC was identified within the *Norme Tecniche di Attuazione* (*NTA*), i.e. technical standards for implementation of the Regional Water Protection Plan (*PRTA in the Italian acronym*), such as instrument for achieving the objectives of protecting and managing water bodies. In particular, article 10 of the NTA highlights how «the Water Protection Plan is implemented through a coordinated action on the part of all accountable institutions [...] by way of negotiated procedures or tools and environmental agreements. [...] In this case, the negotiated programming tools are denominated river or lake contracts». Even in the context of *Piemonte*, the PTR acknowledges the capacity of river or lake contracts to develop synergies with provincial or local land-use planning (article 35, paragraph 3) as well as their characteristic of being negotiated programming tools, correlated to strategic planning processes for the requalification of river basins, oriented to defining, *in itinere*, a common path with all stakeholders in order to promote the integration of the various policies (article 35, paragraph 4). Furthermore, it is noteworthy that from a legal standpoint the Regional Administration identified RC with Negotiated Programming Agreements, under the provisions National Law No. 662/1996 on Rationalization measures of public finance, thus defining these same contracts in a more explicit way than in other Italian Regions. In this sense, the RC is identified as a form of regulation adopted between public bodies, or between an appropriate public entity and one or more parties, whether public or private, for the implementation of several actions, aiming at a single development goal, requiring a comprehensive assessment of the activities for each involved actor is accountable. In addition, the Regional Administration of *Piemonte* was the only local authority to have drafted its own *Regional Guidelines for the implementation of River and Lake Contracts* (Decision of the Regional Council, No. 16/2610 of September 19, 2011).

In the case of the Region of *Emilia Romagna*, the adoption of RC makes reference to the orientations defined by the *Regional Landscape and Territorial Plan* (PPTR in the Italian acronym) and by integrated and experimental projects. The latter have been promoted by Regional Law No. 20/2000, which sanctioned the right of local communities to participate in the preparation of regional and local policies (Montaletti 2011). Contrary to the preceding two regional contexts, in *Emilia Romagna* regional regulations have yet to explicitly acknowledge RC as viable measures for implementing river basin and regional water protection plans. The only reference in this regard can be found in the *Territorial Coordination Plan (PTCP in the Italian acronym) of the Province of Bologna*, which, under article 1.3.5 inserts the so called *River Pacts* amongst the possible tools for implementing the measures of the Provincial Water Protection Plan.

Continuing with the analysis of the regulatory framework of RC, another interesting case is that of the Region of *Puglia* within whose Regional Landscape and Territorial Plan (PTPR), the *Val d'Ofanto River Contract* was identified as the pilot project aimed at creating a local network of experiences of active citizenry so as to raise public awareness among local inhabitants, regarding the value of the landscapes of *Puglia*, as well as to set in motion processes of cooperation and exchange, even within the same communities.

On the supra-regional scale, it is interesting to note how RC are envisaged under most management plans for hydrographic districts in Italy to be prepared by individual River Basin Authorities. For example, the *Management Plan of the Hydrographic District of the Po River* states that the implementation of integrated actions, programmed at the basin and sub-basin levels, can make use of negotiated planning tools, such as river and lake contracts.

An explicit reference to RC is also found in the management plans regarding the hydrographic districts of Sicily, Northern Apennines and Eastern Alps, specifically in the management plan of the *Brenta* and *Bacchiglione* rivers. In the latter case, in particular, the RC for the *Astico-Tesina* sub-basin was acknowledged among the measures necessary to achieve the environmental objectives targeted in the Basin Plan.

In point of fact, the only reference to cross-regional contexts, while not of a legislative nature, thus clearly not regulation yet, is represented by the *National Charter of River Contracts*, presented in 2010, during the *5th National Conference on the River Contracts*, held in Milan Specifically, the National Charter identifies these voluntary agreements as negotiated and participatory planning processes aimed at containing eco-landscape degradation as well as at the requalification of river basin and sub-basin territories (Bastiani 2011). A substantial aspect of this initiative consists in the fact that this propositional document was advanced by the *National Working Group on River Contracts* involving Regions, Provinces, associations among municipalities, the research sector and universities, associations and representatives of commercial enterprises and professions. It intends to be a guiding tool for the dissemination and application of the RC paradigm to various local

Italian contexts. Evidently, the overarching goal of the National Conferences on River Contract is to spur national lawmakers and Regional administrations alike to update current legal frameworks with new legislation specifically dedicated to the regulation and funding of RC.

3.4 Contents and Procedures

In the various European, national and regional contexts, river contract have different regulatory frameworks of reference, objectives, contents and implementation procedures, in the various national and regional contexts. Specifically, in France and in Italy these contractual instruments are characterized by local procedural aspects, as it is illustrated in the following paragraphs.

In France, *Contrat de Rivière* entail technical and financial agreements that define action programs and interventions, generally undertaken over a five-year timespan. As mentioned above, their contents and procedures are regulated by the *Circulaire relative aux Contrats de Rivière et de Baie* of 2004, that classified the relative framework in terms of contents into five thematic sections (*volets*):

- (A) regulatory enforcement measures to counter pollution in order to improve water quality;
- (B1) restoration and renaturation works on riverbanks and beds, coastlines and areas subject to floods; integrated management actions and measures for improving the value of aquatic and marine environments and related landscapes; protection of fish species;
- (B2) works to improve the safety of inhabited areas against the risk of high waters and floods and rising sea levels;
- (B3) actions to improve the availability of water and especially drinking water;
- (C) local promotion and coordination, monitoring and evaluation over the medium and long term of actions and programs in the context of the contractual agreements.

Therefore, although tailored to the different local scenarios, the objectives of each *CdR* must be encapsulated within the above schema, according to the appropriate basin or sub-basin territorial scale, and in compliance to the *SDAGE* and *SAGE* water planning tools, regarding a given basin.

The procedure for drawing up and implementing the terms of the *CdR* advances through preparatory phases for the drafting of a program of actions as prerequisites for their subsequent realization. The implementation process of a *CdR* ends with an evaluation phase of the results achieved, so as to draw overall conclusions on the initiative and eventually extend it with a next contract.

The initial phase consists in mapping out the knowledge base about the interested hydrographic territory, through measures of a *diagnostic de bassin versant*, based on which a *dossier sommaire de candidature* is then drafted. This phase, therefore, takes on the form of an assessment of the state of the pertinent areas by a local agency, such as a *syndicat de bassin*, then sketching out a first draft of objectives and lines of action shared amongst the *partenaires locaux*. In this respect, stakeholders rallying to a territorial consultation initiative represents a crucial facet, able to raise collective awareness regarding the need for integrated water management that is, at the same time, respectful of all the interests involved.

In the second phase, the proponent institution submits the *dossier sommaire de candidatures* to the *comité de bassin*, that validates the proposed *CdR* in terms consistency to the water planning and programming tools already in place for the specific basin or sub-basin.

In the next step, the *Préfet du département* institutes the *comité de rivière* that primarily draws up the *dossier definitive*, providing a more detailed account of the program of actions, the economic framework, the funders and the actors responsible for implementation of the planned measures.

The fourth step is the presentation of the final project to the *comité de bassin* for its validation and final approval. This phase concludes the preparation procedure of the *CdR* and leads to the underwriting of the contract by all partaking public and private entities, as well as by its funding bodies.

After the signature of the contract, follows the launch of the actual implementation phase of the action plan, generally lasting for a five-year period during which the *comité de rivière* constantly monitors the ensuing operations and results.

The conclusion of the contract marks the start of the last phase during which the *comité de rivière* makes an overall assessment and draws up the bottom line of the endeavor. On the grounds of this assessment, the committee may opt for the implementation of a new *CdR*, a *SAGE* (if not yet adoptetd) or another integrated water management solution.

The first two stages constitute the period of émergence of the contract and last approximately 2 years; stages 3, 4 and 5 represent the *élaboration* stages and on average they are concluded within about 4 years from the start; finally, the sixth and the seventh stages represent the period of the *réalisation* of the actions program, which in aggregate add up to an overall process duration averaging 6 years. The nature of the instrument in terms of concertation and participation explains the reason for its ample requirements in terms of time, considering the different phases of negotiated and the complexity of both consultation and implementation activities, entailing consensus among a plurality of politico-institutional and socio-economic actors, whether public or private.

In contrast, in Italy there is no standardized procedures for drawing up and implementing river contracts. In fact, although the concept of integrated water management and a concerted contractual approach are the common denominators across the national context, in each regional realities different methodologies have been adopted for RC development and implementation, often conditioned by their distinctive territorial features and the local regulatory framework of reference. Nonetheless, from an operational standpoint, RC translate into programs of shared activities whose economic sustainability is deemed essential.

In some cases, RC strive to pursue multiple objectives ranging from protection, mitigation and prevention of hydrogeological and flood risk, to improving the value of landscapes, from the sustainable development of tourism to the dissemination of a *water culture* (Bastiani 2011).

Although far from representing two paradigms valid for every Italian administrative and territorial context, both the *National Charter of River Contracts* and the *Piemonte Regional Guidelines for the implementation of River and Lake Contracts* offer two first reference points. On examination, these documents reveal some essential steps to the groundwork, implementation and evaluation procedures regarding programs of actions and interventions.

The first of these phases foresees the activation of the consultation process and the building of a network of motivated local actors, interested in the redevelopment of a watercourse and its territory. Once the politico-institutional and socio-economic scenario within which the RC is to be activated is defined, the second phase follows dealing with identifying objectives, resources and critical aspects, as regards the given context, and with the necessary actions program. In the third phase, the agreement is formalized, whereby all public and private bodies partaking in the initiative underwrite the agreement. Next comes the phase of practical implementation of the program of actions and interventions and, lastly, the concrete results achieved are monitored for the entire duration of the contract.

Throughout all the above stages, cross-territorial promotion initiatives are taken with regard to communication and raising awareness on the part of local communities and other involved stakeholders.

Time-wise, the duration of RC in Italy, unlike many of its European counterparts, is anything but rigidly set and, to this effect, the *National Charter* states that they are to remain in place for as long as the actors underwriting the agreement have the will to accede to it.

3.5 Stakeholder Roles and Participation

In the French context, the *contrats de rivière* have promoted significant consultation processes between various public and private actors, as well as forms of active participation on the part of a range of stakeholders, including in particular the *usagers* (the actual users of water resources) to the benefit of providing them with a greater sense of responsibility (Brun 2010).

On the basis of procedural stages described in the previous paragraph, the institutional profiles and roles of individuals actively taking part in the elaboration and implementation of a *CdR*, can be identified.

Amongst the key actors there are the *syndicats de rivière*, who, for the most part, represent the leading promoters of these agreements. With regard to the action program, the *syndicats* take on the role of *structures porteuses* of the *CdR*, also becoming the protagonists of its implementation, by setting in motion the actual activation process and coordinating the actors involved. Therefore, the *structure porteuse* represents the organism that brings stakeholders together and identifies public and private funders. In addition, it is up to the *structure porteuse* to map out the initial knowledge base regarding the specific hydrographic context and draft the *dossier préliminaire*, and then to submit it for approval on the part of the *Comité de basin*, which represents the decision-making level intermediate between the State and the local governments. In this way the *structure porteuse* outlines the orientations regarding integrated water resources management and related ecosystems, also fostering appropriate consultation and participation measures and procedures.

If the preliminary dossier clears the assessment hurdle of the *comité de bassin*, a *comité de rivière* is then instituted by the *Préfet* of the *department*. The composition of this committee tends towards an equitable representation of the various institutional, social and economic components of the territory of reference. The *comité de rivière* is chaired by a representative of the local institutions and constitutes the arena for consultation amongst all parties partaking in the *CdR*, so as to first draw up the *dossier definitif* and, upon its approval and funding, ensure the implementation of the actions program. With the aim of guarantying proper governance of the RC implementation, the *comité de rivière* consists in three colleges comprising representatives from (I) local authorities and local public bodies, (II) State administrations, (III) *usagers*, associations and organizations. Such a tripartite architecture, in effect, tends to ensure that the principles of territorial consultation be respected not only during each stage leading to the final agreement, but also during the definition of the objectives, contents and the rationale for its actions, especially through the workings of specific sub-committees.

Given the case whereby the territorial area of a *CdR* and that of a *SAGE* overlap, entirely or in part, the competences of the *Commission Locale de l'Eau* (*CLE*) also come into play. In particular, if the two instruments apply to the same geographical area, the *CLE* takes on the role of *comité de rivière*; vice versa if the territorial area of the *CdR* coincides only in part with that of the *SAGE*, the *comité de rivière* constitutes as a special internal sub-committee of the *CLE*. Finally, when the perimeter of the *CdR* is instead greater than that of the *SAGE*, the *CLE* members of that particular territory consequently converge into the make-up of the *comité de rivière*, in representation of the institutions and communities of the area of reference.

During the formulation phase of the *CdR*, the *comité de rivière* performs a thorough evaluation of the envisioned interventions and program of actions, by way of its technical sub-committee, and subsequently submits the final dossier to the *Comité de bassin* for the required approval. If the resulting evaluation proves favorable, the agreement is then formalized and approved so that it may be ratified by the *Préfet du Département*, acting on behalf of the State, and then signed by all the other partners.

Throughout the implementation phase, both the *structure porteuse* and the *comité de rivière* supervise the progress of the *CdR*, aiming at ensuring the thorough consistency of actions, in particular with reference to the scale of the hydrographic basin, and the appropriate integration with all other existing contractual procedures. Precisely, the *structure porteuse* assumes a more executive and operational role, performing interventions according to its specific competences, while coordinating and assisting the actors entrusted with the implementation of the remaining actions. Furthermore, it schedules the meetings of the *comité de rivière* and conducts the necessary activities of territorial animation, consultation, communication and providing information.

In parallel, the *comité de rivière* is poised for the constant monitoring of the headway made according to the terms of the contract, in particular by reviewing the annual reports drafted by the *structure porteuse*, so as to adjust the route of the program of actions, in the course of regular meetings held at least on a yearly basis until the termination date of the RC.

In the overall scenario of the actors involved in each *CdR*, others playing a lead role are clearly the financial partners, such as the State, the *Ministère de l'Environnement*, the *Agences de l'Eau*, the *Régions* and *Départements*, in addition to eventual private funders.

By and large, the ample representation of local bodies, characterizing the preparation and implementation phases of a *CdR*, testifies to the significant bearing that local political representatives have gained over the long evolutionary process of such instruments. To this effect, these representatives are deemed by some authors to be the actual «*principaux artisans des contrats de rivière*» (Brun 2010, 311).

In contrast to other European countries and France in particular, Italy lacks any univocal definition of the different roles of the actors involved in the process of activating and realizing RC. In fact, the current scenario suffers from marked cultural, regulatory and procedural delays, despite, as mentioned above, the contents of the *National Charter of River Contracts* representing common reference keys for many Italian contexts.

In most experiences initiated in Italy, regional and provincial governments have played the role of promoters, coordinators and supervisors of the programs of actions foreseen by RC, on account of (I) their specific institutional competences in the field of water resources management, (II) the subsidiarity principle, and (III) their duty to practically integrate RC into the instruments of urban and territorial planning. Normally, regional and provincial administrations map out and draft the initial knowledge base regarding the hydrographic territories of interest and the actions programs, or they alternatively commission their drafting, to other institutional, scientific and technical organisms. The same administrations, then supervise the realization of the RC, ensuring that the commitments taken on by the subscribing parties are met and that all actors operate in a coordinated and synergistic manner (Bastiani 2011).

In the current scenario, even municipalities take upon themselves, or at least frequently share, the role of promoters and actuators, both through their own initiative and in the form of partnerships with other municipalities. Similarly, the local *consorzi di bonifica* (irrigation consortia), as associations between municipalities, park authorities and environmental organizations, may promote and enable the preliminary processes for stipulating new RC.

In some cases, the entity in charge relies on the technical and scientific support of *Regional Environmental Protection Agencies* (*ARPA in the Italian acronym*), River Basin Authorities, universities and other research institutions, particularly while mapping out the knowledge bases, necessary to define actions programs and related interventions.

In those regions with the most seasoned track records of RC, such as *Lombardia* and *Piemonte*, the drawing up of the contract and its implementation are delegated to a control room, usually comprising representatives of local authorities, River Basin Authorities, park authorities, *ARPA* and other main stakeholders involved. This executive body, therefore, fulfills both a policy-making and a coordination role, and represents the seat where local issues are addressed in technical and scientific detail, objectives are focused and defined, and the activities to be included in the action plan are aptly configured.

The control room normally avails itself of the support of a technical secretariat, that is a panel made up of the local technicians of reference, external consultants and different domains experts. To the control room are delegated the functions of coordinating the various phases and executive activities of the RC action program.

In some cases, the seat for consultation activities and agreement management is represented by the *Basin Assembly*, a body composed of institutions and agencies directly or indirectly having administrative and management competences over the same basin. Representatives of local communities may also take part in the Basin Assembly, in various forms.

In some Italian regions, it is also becoming customary to set up a technical panel for a coordination at regional level, composed of representatives of local authorities, acting as an observatory on the RC and for the opportune integration amongst the various programs of actions.

In the overall view of the actors involved in the implementation of RC, a crucial role is also played by funders, which in most cases are represented by the corresponding regional administrations, Basin Authorities and local consortia. In this regard, it is noteworthy that in Italy public financial contributions towards the implementation of RC considerably outweigh all private contributions combined.

In the light of the above, the current scenario of RC in Italy is characterized by regional governments as protagonists, both in the promotion and in the actual implementation phases, especially in those cases in which the contract involves territories spanning boundaries of multiple provinces and/or regions.

3.6 Experiences of River Contracts

In order to gain insight into the evolutionary pathway of river contracts in France and Italy, it is useful to outline a brief overview of the main experiences launched, whether fully implemented or still underway. Given that these two national contexts currently differ in terms of organizational and structural aspects, in France RC will be more cursorily delineated so as to focus greater attention on Italian cases. As a matter of fact, for quite some time France has maintained its GEST'EAU web portal (www.gesteau.eaufrance.fr) that provides constantly updated data and statistics on the *CdR*, and all other water resources management and planning tools including

SDAGEs and *SAGE*, as well as on other *contrats de milieu*. In Italy, however, no such source of information is yet available, which is the rationale underlying the choice of a more detailed overview of the Italian RC experiences.

In France, the first three *CdR*, signed in 1983, were for the *Thur* river, pertaining to the *Rhin-Meuse* Basin (*Région Alsace*), for the *Loiret* river (*Région Centre*) and *Trieux* river (*Région Bretagne*), with the latter two both belonging to the *Loire-Bretagne* basin. Since then, the diffusion of *CdR* has undergone significant acceleration: in 1984, there were a total of 9 cases, in 1990 there were 25, while in 2004 the number reached 179. To date, altogether 269 contracts at varying levels of advancement have been stipulated throughout the country. Of the total, 148 have already been completed, another 67, albeit already signed, are still being implemented, 46 are being processed and another 8 contracts are still at the emerging phase (Fig. 3.3). In addition, 29 of the total 269 contracts constitute cross-border projects in cooperation with Switzerland, Belgium, Spain and Italy, 3 involve the overseas French regions of Guadeloupe and Martinique, whereas 54 resulted from institutional interactions between two or more Regions.

On closer examination, the *Carte de situation des contrats de milieu* (www.gesteau.eaufrance.fr) reveals an uneven distribution of *CdR* on the national territory, both amongst *Régions* as within river basins, where there are consistent differences mainly reflecting the diversity in national and regional policies in matters of financial support.

With specific reference to the number of *CdR* stipulated per regional area, it can be inferred how most of them are concentrated in the *Région Rhône-Alpes* (48 %), *Provence-Alpes-Côte d'Azur* (15 %) and *Languedoc-Roussillon* (12 %). In addition, a significant number of RC is comprised in the *Région Midi-Pyrénées* (9 %), attesting to its determination in protecting its waterways of elevated environmental quality and, thus in stimulating a greater sense of responsibility on the part of the local political representatives. Likewise, equally significant numbers are found in the *Région Bourgogne* (7 %) and *Auvergne* (7 %) as well, due to their proximity to the *Région Rhône-Alpes*.

On further examination of the French scenario, proceeding from North to South and from West to East, we can mention some of the most representative *Régions* as far as the diffusion process of *CdR* is concerned.

In *Région Bourgogne*, 41 % of the total surface area is comprised in 19 *CdR* already initiated and underway. As for the total number of ongoing initiatives, with regard to the *Vouge*, *Armançon*, *Ouche* and *Tille* rivers, the *CdR* are currently being integrated with their respective *Schémas d'Aménagement et Gestion des Eaux* (*SAGE*), which are already approved in part, while in other cases they are still in progress.

The *Région Auvergne* numbers 19 active *CdR*, regarding approximately 47 % of its territory. A particularly noteworthy aspect, with reference to the resources made available for the realization of *CdR*, is represented by the fact that 30 % of the total funding for each contract submitted is covered by the Regional administration. This quota of funding, combined with the resources put forth by the *Agence de l'Eau* and other public bodies, amounts to up to 80 % of the total costs for projects.

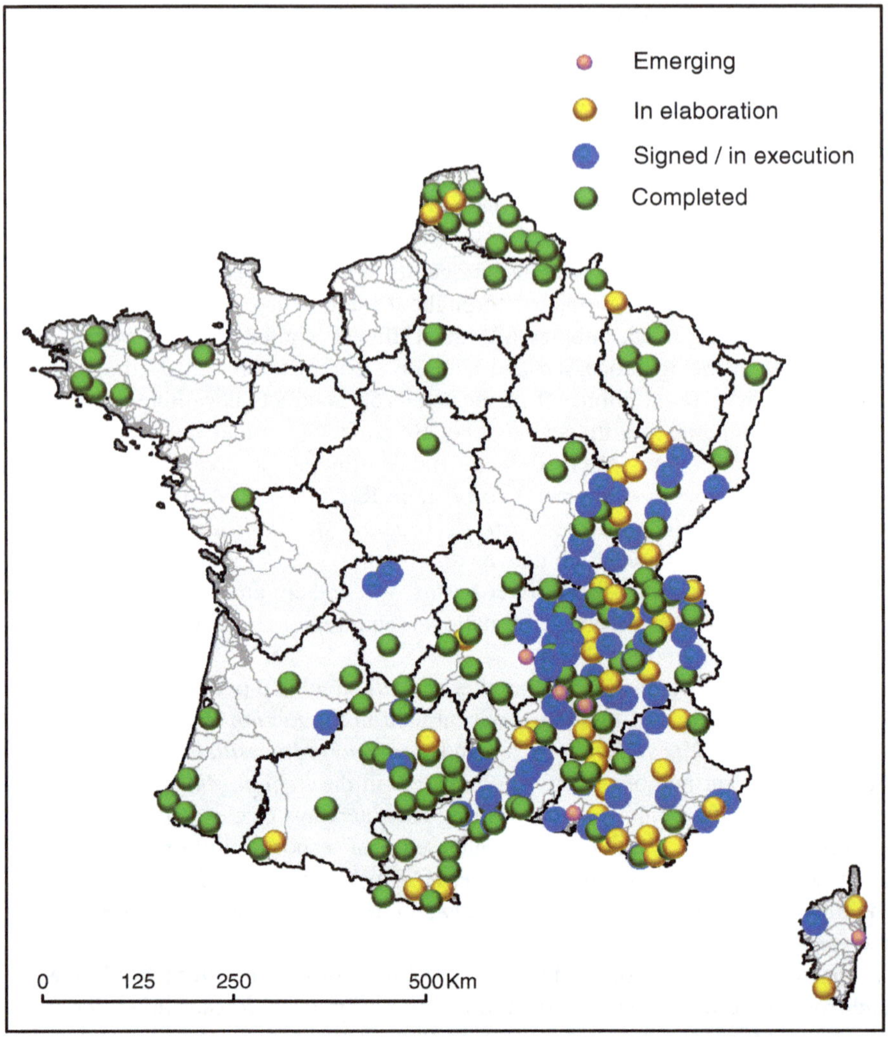

Fig. 3.3 Distribution map of *contrats de rivière* with related levels of advancement (data from GEST'EAU web portal)

The *Région Rhône-Alpes* is unmistakably the most dynamic, considering that for 99 % of its territory there are currently *CdR* already in force. Ever since its first contractual agreement, signed in 1984 with regard to the *Ardeche Claire* river, the *Région Rhône-Alpes* has launched a host of other initiatives, thanks to a vigorous institutional collaboration with the *Agence de l'Eau Rhône Méditerranée-Corse*, especially as regards the so-called second generation of *CdR*, following the adoption of the *Loi sur l'Eau* of 1992.

Of note, in 2000 the Regional Administration conducted a comprehensive assessment of *CdR* in progress at the regional level, in concert with the *Direction Régionale de l'Environnement* (*DIREN*) and the *Agence de l'Eau*. The latter appraisal occasioned a process of reflection on some of the critical aspects of these agreements, on the actual results in terms of water quality and river environments, on the possibility of renewing such procedures, as well as on the reasons underlying the limited dissemination of *SAGE* in this *Région*. In particular, this analysis highlighted the need to boost *ex ante* consultation processes and ensure more effective integration with other water management initiatives and urban and territorial planning tools. In addition, an analysis of the procedures in place at that time highlighted the importance of *Etudes d'opportunité* and *Etude de bilan et perspective*, so as to verify and ascertain the motivations behind the financial commitments required by *CdR*, advocating such studies as milestones for the implementation of new initiatives. It is also noteworthy that, with regard to the *Région Rhône-Alpes*, by and large the *CdR* have represented convenient alternatives to *SAGE*, thanks to their greater flexibility and adaptability to individual local contexts and, thus, their relative ease of implementation in procedural, financial and temporal terms.

In the *Région Midi-Pyrénées* there are also numerous examples of *CdR*, which, along with *SAGE* and other water management plans, represent the main devices put in place for purposes of safeguarding, improving the quality status and the conscientious use of water resources. To date, there are 25 *CdR* in force throughout the Region. It should be noted that in this specific case, the *Conseil Régional Midi-Pyrénées* finances programs of action included in the *CdR* with a contribution in the measure of about 40 % of total project costs.

The *Région Languedoc-Roussillon* counts 32 *CdR*, either fully activated or at varying phases of implementation, covering 66 % of the entire regional territory. Among the cases arising in this setting, that of the *Contrat de Rivière de l'Orb* reveals itself to be extremely interesting especially in view of the assortment of seats for consultation and governance, coming into play from the very onset of the emerging contract (Richard 2005). In this case, an initial contract signed in 1996, and mainly aimed at measures for improving water quality and flood protection, was followed in 2006 by a second contract that has been completed in 2010, with the achievement of interesting results. In particular, it is remarkable that over the entire 14-year life-cycle of the *CdR*, in the seats of consultation and governance, i.e. the *comité de rivière*, the *syndicat mixte de la vallée de l'Orb* and three *comités consultatifs*, participation kept mounting, also to accommodate a host of private players active in the area.

In the *Région Provence Alpes Côte d'Azur* the *CdR* represent, along with the *SAGE* and the *Schémas d'aménagement de bassin versant*, the tools put in place by the *syndicats de rivière* for integrated water resources management at the river basin scale. To date, the 41 *CdR* launched, involve about 70 % of the regional territory. As of 1999 the *Agence pour l'Environnement Régional Provence Alpes Côte d'Azur*, the *Agence de l'Eau Rhône-Méditerranée-Corse* and the *Direction Régionale de l'Environnement, de l'Aménagement et du Logement* (*DREAL*) have created a regional network of subjects managing water environments, so as to promote better coordination of all public and private actors operating in the sector. In 2010 the Regional Administration drew up and published a *Guide pour une gestion durable des milieux aquatiques*, in order to assist the various management bodies in integrating the sustainable development processes into the framework of *contrats de milieu* and, consequently, *CdR*. In the wake of this initiative, the first *CdR durable* was launched in early 2011 with regard to the *Bleone* river and its tributaries. It is to pointed out that on this occasion the main themes of sustainable development (climate change, biodiversity, quality of life, social cohesion, sustainable consumption and production) were tailored to meet the specific needs of the *Contrat de rivière Bleone*.

Since 2004, in parallel to *CdR*, numerous ongoing efforts to draw up *contrats de canal* have been underway. The latter are oriented towards an integrated management of water channels, encompassing their various uses, the quality of groundwater, rainwater harvesting and the cultural, social and economic values typical of such environments, whose connotations are characteristically twofold, i.e. natural and anthropic (Chémery and Luczyszyn 2011).

In Italy, the river contract scenario is, on the whole, less advanced and evolved in comparison to the French framework.

The analysis of Italian cases shows that after a first series of RC, launched as of 2004 in the Region *Lombardia*, a second season of initiatives followed, upon receiving a distinct boost from the presentation of the *National Charter of River Contracts* during the *5th National Table on River Contracts*, held in Milan in 2010. To date, 9 Italian regional administrations have adopted and signed this programmatic document, and another 5 are in the process of accession, while numerous examples of RC are being implemented in 19 out of the 20 Regions.

From North to South, from West to East, the analysis of the Italian scenario offers a variegated and differentiated picture of programs of actions and a range of levels of integration with other water management tools and urban and territorial planning instruments, as well illustrated by numerous contributions offered by many expert in occasion of *10th National Table on River Contracts* held in Milan at *Regione Lombardia*, from 15 to 16 October 2015. To date, altogether 116 contracts at varying levels of advancement have been stipulated throughout the country. Of the total, 11, albeit already signed, are still being implemented, 53 are being processed and 52 contracts are still at the emerging phase (Fig. 3.4).

In the Region of *Valle d'Aosta*, the reflections and experiences of RC have stemmed from the *Alcotra Eau Concert* EU project, financed through a Cooperation Program between Italy and France (2007–2013), aiming at consolidating a shared knowledge framework and pursuing a program of pilot efforts to improve participated water management policies in support of river ecosystems. In particular, the project activities were geared to activate a RC for the *Dora Baltea* river, based on the experience already gained from the neighboring Region *Piemonte*, on the one hand, and from the *Guidelines for River and Lake Contracts* authored by this same Regional Administration, on the other. The activities of the project, completed in February of 2015, were conducted in synergy with the *Départements* of *Savoie* and

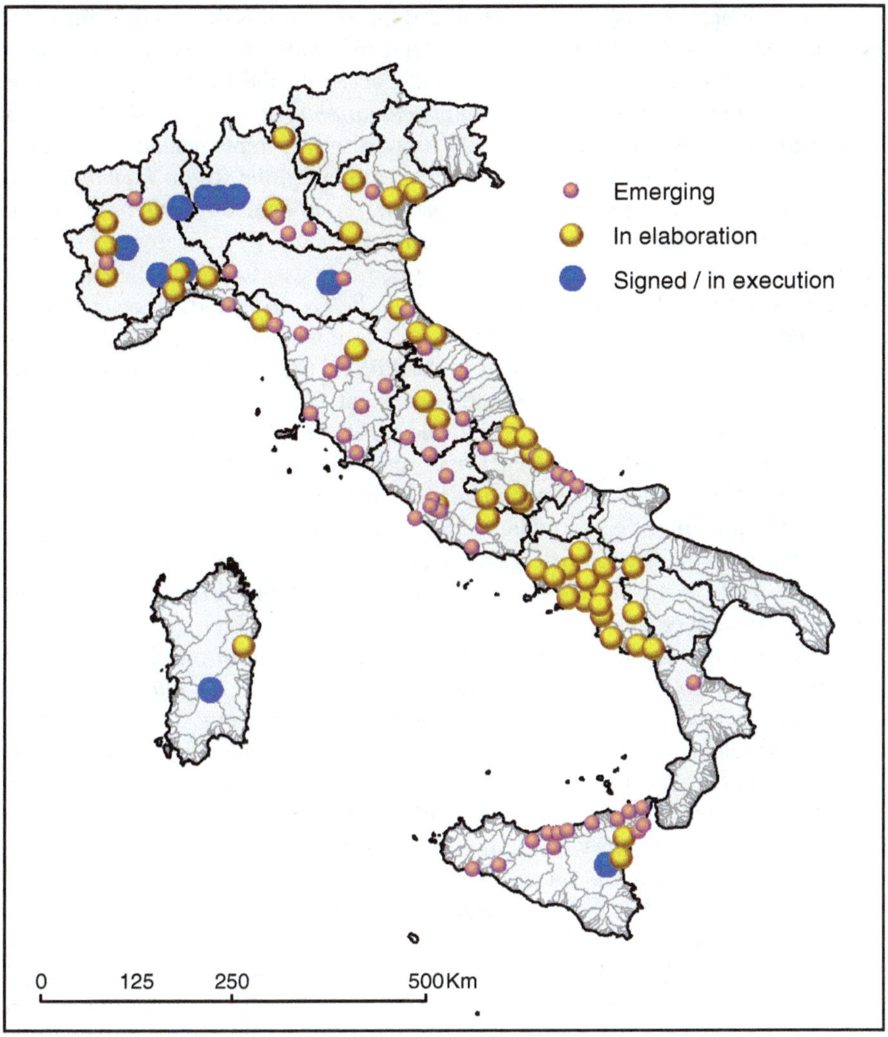

Fig. 3.4 Distribution map of river contracts with related levels of advancement

Haute-Savoie (*Région Rhône-Alpes*) and *Hautes-Alpes*, the *Alpes-de-Haute-Provence* and the *Alpes-Maritimes* (*Région Provence-Alpes-Côte d'Azur*), in order to systematize respective expertise in matters of integrated water management and RC.

In Region *Piemonte*, RC were identified as negotiated and participatory programming tools, suitable for the implementation of the *Regional Water Protection Plan* (*PRTA*). The RC was also included within the *Regional Landscape and Territorial Plan* (*PPTR in the Italian acronym*), as an operational instrument aimed at the improving the quality of river basins as well as the integration of the various water management tools (Clemente et al. 2011). Also in this perspective, thanks to research done in collaboration with the Inter-University Department of Territorial Studies at the Polytechnic and the University of Turin, the Region *Piemonte* prepared the *Regional Guidelines for the Implementation of River and Lake Contracts,* mentioned above, as a means of technical support for the implementation of present and future sectoral initiatives. A fact worth mentioning, the RC in *Piemonte* have concerned river basins presenting considerable environmental challenges, due to their being under strain mainly from anthropic pressures. Specifically, deliberation on the use and application of this kind of agreements was fueled by the experience accruing from four pilot basins, namely *Agogna*, *Belbo*, *Orba* and *Sangone*. In more general terms, the prospects for the Regional Administration are to apply these RC to all 34 hydrographic areas indicated in the PRTA, transforming these negotiated tools from experimental initiatives into permanent forms of participated management of fluvial contexts (Governa and Toldo 2011). To date within the territory of Region *Piemonte* there have been 10 river and two lake contracts, either fully in effect or at various stages of implementation, based on various institutional partnerships. In particular, the Bormida and Scrivia RC are both characterized by joint cooperation with the Region of *Liguria*, whereas the *Dora Baltea RC* involves the Region of *Valle d'Aosta* and the French partners of the *Alcotra Eau Concert* project, as mentioned above.

The Region *Lombardia* constitutes undoubtedly a frame of reference for RC in Italy. Indeed, in 2004 the Regional Administration initiated *the Olona -Bozzente-Lura River Contract* by means of an *Accordo Quadro di Sviluppo Territoriale*, i.e. a territorial development framework agreement. The foundations of this experience were rooted in the reflections born out of the participation of Region *Lombardia*, in 2000, in the project entitled *NetWet 2 Water Telematic Platform* (founded by INTERREG IIIB CADSES Community Initiative Program 2000-2006), envisaging a pilot RC for the *Lambro-Seveso-Olona* basins, characterized by high environmental and hydrogeological risk (Bastiani 2011; Clerici et al. 2011). In the wake of that initial experience, the *Seveso River Contract* and a *Memorandum of Understanding for the Mella River* were signed in 2006. Between 2006 and 2008, action plans were commenced with the aim of re-qualifying the *Mincio* River and *Oglio* sub-basins, lying within the two homonymous regional parks. In 2007, contracts for the Norther *Adda* River and the *Iseo* Lake embarked on their preliminary phases of activation and, finally, in 2011 the *Lambro* River Contract was started (Bartoli and Perlini 2011). Generally speaking, an analysis of the experiences carried out in *Lombardia* reveals how most RC have been activated in hydrographic sub-basins that result, in large measure, heavily compromised in terms of the integrity of their ecosystems and ecology (Bastiani 2011; Clerici et al. 2011).

In Region of *Trentino Alto Adige* the RC activated are related to the sizeable river basins of the *Adige* and, indirectly, the Po rivers. Specifically, as of 2013 the *Basin Authorities of the Adige and Northern Adriatic Rivers*, responsible for the *Hydrographic District of the Eastern Alps*, have been engaged in implementing the River *Contract of the mouth of the Po river*, whose Memorandum of Understanding was initialed in April of 2015. In this regional context, the *Master Plan of the Sarca River Park* is also underway in order to consolidate an institutional cooperation process between the Autonomous Province of Trento and the municipalities of the Giudicarie valleys and the Lower Sarca—Valley of Lakes. The action plan, although embracing the general river-contract paradigm, makes customized adjustments to the negotiated approach also to meet their specific objectives pertaining to water resources management and fluvial ecosystems.

The Administration of the Region *Veneto*, by subscribing to the National Charter of River Contracts in 2013, embarked on a path of promotion and implementation of RC as a model for territorial governance and integrated water resources management. At the foundation of this stance there is an acknowledged need for institutional synergy so as to achieve the objectives of the hydrogeological safety and sustainable urban and territorial planning, as well as to protect and enhance the quality status of water resources and related environments. In addition, in 2015 the Regional government approved the institution of a Regional Table for the Coordination of River Contracts, comprising representatives of various regional departments and the three Basin Authorities of the Po river, the upper Adriatic area and the Adige river, as well as other local stakeholders. In *Veneto*, the first RC was promoted and initiated by the Province of *Vicenza* in 2010 with regard to the *Astico-Tesina* stream, in order to restore and re-qualify its fluvial and environmental status. In the same year, the Municipality of *Silea*, in the province of *Treviso*, launched an initiative *Towards the Melma and Nerbon River Contract* with the ultimate aim of curbing urbanization and anthropic exploitation of the territory within the territorial area comprised by the two streams (Bastiani 2011). In 2012, on the proposal of a local irrigation consortium, the declaration of intent regarding the RC of the mouth of the Po river was signed on behalf of territories at the mouths of the major coastal rivers, *Brenta*, *Adige* and the *Eastern Po*, as well as the various branches of the *Po*, besides the related areas on the Adriatic seashore. To date, the mapping out of the knowledge base and other preparatory steps for the signing of the Contract for the Mouth of the Po River are still underway. In 2013, another local irrigation consortium promoted the *River Contract of Marzenego-Osellino*, involving the Provinces of Treviso, Padua and Venice. With funding from the Region Administration of Veneto, the knowledge base and a concerted program of actions are currently being defined for this RC. Finally, three other experiences in the making are the *River Contract Brenta*, for which the public consultation process was initiated in December of 2014, the *River Contract for the Meolo-Vallio-Musestre*, the Memorandum of Understanding for which was initialed in August of 2015, and the *River Contract for the Adige Euganeo*, under the auspices of a homonymous local irrigation consortium, as of 2014. The corresponding declaration of intent, entitled *Towards the River Contract of the Adige Euganeo*, was undersigned in May 2015 to actively involve all stakeholders in the territory comprised between the rivers Fratta-Feria-Gorzone, Adige, Brenta, Bacchiglione and the lagoon of Venice.

In the regional context of *Liguria*, to date there are 5 existing RC, and three of them represent initiatives shared Region of *Piemonte*. In 2012 the Memorandum of

Understanding for the *River Contract of the Erro* stream basin was undersigned and in 2014 the Provinces of Alessandria (*Piemonte*) and Savona (*Liguria*) adopted its plan of actions, subsequently submitted to the necessary Strategic Environmental Assessment procedure. In the case of the *River Contract for the Scrivia stream*, regarding an area overlapping Liguria, Piedmont and Lombardy, the Memorandum of Understanding was signed in March of 2013 and was followed through to the approval of the actions program, on the part of the Provinces concerned, in 2014. Another interesting example of RC was the one activated for the Bormida Sub-basin, overlapping the Provinces of Alessandria, Asti, Cuneo (*Piemonte*) and Savona (*Liguria*), whose Memorandum of Understanding was initialed in 2013. In this case, the promoter institution was the Province of Alessandria and the partnership includes, besides the above mentioned provinces, 123 municipalities, for a total extension of 2620 sq km of impacted territory. The preliminary procedures relative to the contracts for the Magra and Entella rivers are currently being defined. In particular, in the case of the Magra river, the need to activate a RC also became more pressing following recent floods that hit many of the inhabited areas along its course.

In the Region of *Emilia Romagna*, RC are named as river pacts and are characterized by participated aspects whereby the river-territory-landscape relationships are dealed with (Montaletti 2011; *Pizziolo and Micarellli* 2011). These river pacts are organic to the orientations defined in the (I) Regional Landscape and Territorial Plan, (II) Regional Urban Planning Law No. 20 of 2000 with particular regard to «projects of an integrated nature and experimental character» (Montaletti 2011, 315), and (III) the River Po Basin District Management Plan, approved in 2010. Specifically, the first river pact was activated in 2006 with regard to the basin of the Samoggia-Lavino rivers and was characterized from its onset by an active involvement of local players and communities (Bastiani 2011; Montaletti 2011). Nevertheless, the Memorandum of Understanding of the this RC has yet to be initialed, as is the case for most other experiences in the Region of Emilia Romagna. Indeed, of the 7 RC activated as of 2006–2007, the only one actually being implemented refers to the Middle Panaro river, signed in 2012, with next in line being the one regarding the Marecchia river, whose Memorandum of Understanding was initialed in 2014. For the latter case, it is interesting to note the inclusion of the RC within the framework of a more general strategic planning process regarding vast areas that, starting from the recognition of the river as a unifying element, aims to promote an integrated and participated management of landscape and environmental resources.

For the context of Region *Toscana*, it can be observed that the driving force for dissemination and implementation of RC is supported, on the one hand, by the authorities closely tied to the *Management Plan of the Northern Apennines Basin District* and, on the other, by some spontaneous grassroots movements that advocate novel, bottom-up processes for shaping water resources management policies (Bastiani 2011). In either case, it can be noticed how the participatory dimension, on the whole, has acquired significant weight and, at the same time, how RC have been regarded as tangible tools for implementing *River Basin Management Plans*

(Bonamini and Brugioni 2011). A significant experience is that one undertaken for the *Valdarno Empolese*, started in the mid 1990s, from a lengthy and multi-faceted process of rediscovery of the socio-economic and culture values of this fluvial territory, where local players are currently working together to build a shared planning scenario, with a view to more effective coordination of sectoral planning policies and sustainable territorial development. That trajectory led first to the definition of the *Declaration for the Arno* followed by the *River Park Master Plan*, testifying in this sense to the role of catalyst of the project, played by the *River Contract of the Valdarno Empolese* (Bastiani 2011; Maganghi 2011). Even after the accession of the Regional Adiministartion, in December of 2014 to the *National Charter of River Contracts*, other initiatives have been promoted by local authorities and 9 new RC are currently being activated, relative to the *Albegna, Canale Maestro della Chiana, Carrione, Cornia, Egola, Ombrone Grossetano, deOmbrone Pistoiese, Pesa* and *Serchio* river territories.

The Region *Umbria* is characterized by some interesting experiences regarding RC, in part related to the implementation of the program of actions defined in the *Territorial Strategic Plan for the sustainable development of the regional territories*, approved in 2008 by the Regional Administration. In the Provinces of *Perugia* and *Terni*, initiatives aimed at participated requalification of river environments were created in occasion of the *Local Agenda 21 Forum*, the movement that sprang up following the United Nations Conference on Environment and Development held in Rio de Janeiro in 1992 (Bastiani 2011; Ciarabelli 2011). The activation process concerning the *River Contract of the Tiber river and Upper Umbria* area stemmed from the *Tiber Project* comprised within the *Regional Landscape Plan* approved in 2009 and the *Territorial Strategic Plan* approved in 2011, in view of the Tiber River identity-shaping heritage on behalf of local communities. Starting from several projects and plans for environmental management measures, financed by *European Regional Development Funds* between 2007 and 2013, a preliminary overview of the knowledge framework regarding the *Upper Tiber* sub-basin was consolidated, in order to discern the specifics of the relevant environmental issues, the sectors for priority interventions and the existing relationships between given causes and their effects in terms of environmental degradation and hydrogeological risk. Following this groundwork, the activation process of the River Contract of the Tiber River and Upper Umbria area was set in motion in 2011 via a concerted effort on the part of the Municipalities of *Perugia* and *Umbertide* (Bastiani 2011; Martini et al. 2011). Since 2010, the Province of *Terni* has promoted a participated course of action to delineate the *River Contract for the lower Nera River Valley*, also in line with the orientations of the *Provincial Territorial Coordination Plan*, regarding the stretch of river comprised between Terni and the confluence of the Nera and Tiber rivers (Venti and La Pegna 2011). As part of this initiative, the RC has been identified as an instrument of territorial governance policies and participated processes based on new models of interrelationship between local communities and their territories of reference. So as to improve the ecological status of the Clitunno river hydrographic basin, the Province of *Perugia*, the Regional Agencie for the Protection of Environment and several municipalities have initialed an institutional protocol wherein the RC was viewed as the most appropriate operational approach for achieving the specific project objectives. By way of this cooperative institutional effort, in 2013 the *Irrigation Consortium of Umbria* endorsed a RC for the *Clitunno*, *Topino* and *Marroggia* rivers. Other ongoing initiatives in the Regional context of Umbria concern the Upper Valley of the Nera River, the Marroggia-Teverone-Timia fluvial system, as well as the Paglia river. It is also worth mentioning that the Regional Administration adhered to the National Charter of River Contracts in February of 2014 and that presently, in parallel to the implementation of RC, is working on the activation of the so called *landscape contracts*, i.e. agreements analogous to the same RC.

Also the Region *Marche* adhered in January of 2015 to the National Charter, having identified such instruments as a form of participated strategic planning for environmental requalification and hydraulic risk reduction suitable tools. At present, there are two RC being launched for the *Foglia* river, as part of the *Municipal Strategic Plan of the Municipality of Pesaro*, and that one for the *Esino* river, promoted by the managing body of the homonymous local Natural Reserve. In both cases, however, the Memoranda of Understanding between the institutional counterparts still await signing.

In the Region of *Sardegna* there are currently two active RC experiences regarding the river basins of the *Flumini Mannu* and *Cedrino* rivers, respectively. In the former, the procedural pathway was embarked upon in 2009, in the wake of episodes of hydrogeological upheaval. In particular, it is interesting to note that, although the relevant procedural measures have met up with various obstacles and, to date, are for the meantime suspended, the *River Contract Flumini Mannu* finds mention within the same *Provincial Urban Planning Scheme of the Medio Campidano* (Bastiani 2011; Bandinu et al. 2011). In the second case, the Memorandum of Understanding of the *River Contract Cedrino* has been fostered by the Union of Municipalities of the *Valle del Cedrino* and initialed in 2010 in order to define and implement a system of integrated interventions, aimed at water-quality status restoration and at enhancing the value the fluvial territory of reference.

The context Region *Lazio* offers other interesting examples of RC, the activation procedures of which have all been fostered in recent years. Specifically, the Province of *Frosinone* has promoted the *River Contract for the basin of the Cosa River*, located in the middle regional area. The Memorandum of Understanding of that RC was initialed in February 2011 with the objective of establishing an integrated action plan for the protection, requalification and enhancement of the quality status of water resources and related environments. The preliminary phases of the *River Contract of the Aniene Valley* were similarly initiated in 2011, with the aim of protecting and improving the fluvial environment and its accessibility, in manners that are to be agreed upon and coordinated amongst all actors present in its hydrographic basin, with particular regard to the riverside belts and to the planning orientations adopted at the local scale by each municipal administration. Yet another significant experience, given the basin concerned, is represented by the *River Contract of the Mid Valley of the Tiber River* which involves the territories of several municipalities localized just North of Rome. In 2015, a declaration of intent was initialed by the interested municipalities in order to adopt shared strategies and policies for hydrogeological risk prevention, protection of the Tiber fluvial system, enhancement of its environmental resources and to promote local development. Besides the Tiber river experience, another of considerable interest is the initiative launched in late 2010 by *Legambiente*, a national environmentalist association, for the purpose of drafting a river pact regarding the *Farfa* river. This experience explicitly strives to create needed synergy amongst the Basin Authority, local institutions and the citizenry, but also associations, technicians and farmers, in order to reverse the current trend of uncontrolled exploitation and degradation of the fluvial system, while in turn proposing a more sensible approach to land-use management. It is noteworthy that in 2014 the Region *Lazio* adhered to the *National Charter of River Contracts* while, at the same time, launching an initiative to define a unitary methodological reference framework so as to be able to systematize river, lake and landscape contracts, including their respective orientations and objectives. It need be mentioned that the Basin Authority of the Tiber River has for some time identified RC as participatory tools useful for updating the *Water Resources Management Plan* of 2015 and for drafting the *Flood Risk Assessment and Management Plan* of 2015. Other initiatives for implementing RC are currently being commenced in the Region *Lazio* and, specifically, they concern the *Almone* river, localized in the southernmost districts of the metropolitan area of Rome, the *Sacco* basin, in central regional area, and the land reclamation canals situated in the Province of *Latina*, in the southern part of the region.

In the context of the Region *Abruzzo*, the first experience of RC started in 2010 in connection with the *Vibrata* river basin. This initiative, promoted by the Province of Teramo, was included among the pilot projects of the *Triennial Regional Program for Environmental Protection and Restoration* approved in 2004. Starting in 2014, the implementation process for RC in this regional context has received new impetus following the creation of a special *Department for River Contracts* established by the Regional Administration, the organization of the first *River Contract Regional Assembly* and the approval of the *Regional Guidelines on River Contracts*. Currently, there are a total of 12 experiences in different phases of progress, whereas only three are at an advanced stage, to wit that one regarding the *Tavo*, *Fino* and *Saline* river systems, that one for the Valley *Sagittario* and Lake of *Scanno*, and, finally, the other one *River Contract Tordino*. With respect to the *Tavo*, *Fino* and *Saline* rivers, in 2014 the mayors from the interested municipalities signed the declaration of intent for a specific RC, while the various technical meetings between stakeholders are still underway so as to define objectives and contents for the next agreement. As regards the *River Contract for the Valley of Sagittario and Lake of Scanno*, whose declaration of intent was signed in December 2014 by the mayors of the concerned municipalities, it is to be co-funded by the Provincial Administration and by a group of private stakeholders. Finally, in the case of the *Tordino River Contract*, inserted in the *Provincial Strategic Plan for Environmental Sustainability of Province of Teramo*, the first action program has already been adopted. Furthermore, in the context of Region *Abruzzo* other RC are being activated, namely for the basin of the rivers *Sangro*, *Pescara*,

Vomano-Mavone and *Trigno-Sinello* (the latter in collaboration with the Region of Molise), in addition to a specific contract for the Lake of *Campotosto*. Finally, since 2014 the *Contract for the mouth of the Alento river* has been undergoing activation, while even more recently the plan of action regarding the *River and Landscape Contract for the Valley of the Liri river* has been activated.

In the Region of *Molise*, the sole existing experience, which is actually still being activated, concerns the *River Contract for the Trigno* river, undertaken in collaboration with the Region of *Abruzzo*.

In the context of the Region of *Campania* as well, the RC has been identified by the Regional Administration as a tool to improve the management of water resources at the regional level, while promoting the coordination amongst institutions, associations and stakeholders alike. Consequently, RC were incorporated into the *Regional Water Protection Plan* of 2009, and in 2013 the Region of Campania subscribed to the *National Charter of River Contract*. Currently there are 15 experiences initiated concerning the hydrographic systems of *Volturno, Regi Lagni, Isclero, Calore Irpino, Sabato, Tammaro, Sarno, Ufita, Tanagro, Bussento, Alento, Tusciano, Sarno, Sele* and *Ofanto*, the latter of which promoted in collaboration with the Regions of *Puglia* and *Basilicata*.

The latter Region was one of the first to adhere in 2012 to the *National Charter*, however, to date, few initiatives of RC have actually come to fruition. The only ones that are already underway are the *Val d'Ofanto River Contract*, signed in May of 2014, and another initiative for the definition of an integrated program of actions for the protection and enhancement of the quality status of the *Noce* river basin. With reference to the latter, the initial stage of the launch of the RC occasioned the subscription to a *Declaration for the Noce River*, in 2009, on the part of 1000 signers among the citizenry of the concerned territory (Gerardi and Di Fazio 2011). At present, the relative plan of actions is still in the making through a participated approach fostered by the River Basin Authority, various local institutions and territorial stakeholders.

In the context of Region *Calabria*, the only case of a RC being launched, to date, concerns the *Crati* river, whose activation is at the preliminary stages of local consultation. Worth mentioning, however, is the interesting note that in 2015 the Regional Administration proposed amending the current *Regional Urban and Territorial Planning Law* with an article expressly dedicated to rivers.

In Region of *Puglia* the regard for RC arose following the launch of the *Val d'Ofanto River Contract*, an interregional program agreement signed in May of 2014, intended to promote the integrated and sustainable development of those valley territories, comprised between the Regions of *Campania*, *Basilicata*, and *Puglia* (Bastiani 2011; Iacoviello 2011). In particular, this experience derived from a participated procedure that, in the first instance, entailed the interaction of the regional administration, local institutions and farmers so as to institute the *Ofanto River Protected Natural Area*. To date, the project of the *Val d'Ofanto River Contract* represents one of the most significant experiences of RC concerning an extensive river basin overlapping several regional territories, as better detailed in the Chap. 4 dedicated to case studies.

In the Region *Sicilia* the first RC experience was launched in 2008 by the *Park Authority of the Alcantara River*, by signing a preliminary Memorandum of Understanding, included in the *Regional Water Protection Plan* (Castellana et al. 2011). This experience, albeit suspended to date, represented a reference point for the *Management Plan of the Basin District of Sicilia*, approved in 2009, in particular for those activities concerning the guidelines for drafting and activating RC that would serve as tools for implementing the same management plan. In April 2012, some local institutions and stakeholders of the Province of *Catania* signed the Memorandum of Understanding for the *River Pact for the Simeto*, developed on the base of study and research activities initiated at the end of 2010 by the *Department of Urban and Territorial Planning of the University of Catania*. This declaration of intent, aimed at developing a concrete institutional cooperation, has lead to the signing of the *River Pact of the Simeto*, in May 2015, participated by the relevant local institutions, the University of *Catania*, a local irrigation consortium, as well as by other local actors. Many other initiatives are cropping up almost over the entire Region of Sicilia, especially via institutional partnerships between municipalities, even though not always technically belonging to the same sub-basin. The main reason for the spread of such initiatives lies in the dispositions outlined in the *Regional Flood Risk Management Plan*, pending approval on the part of the Regional Administration. The Memorandum of Understanding have already been signed for other next RC namely *Pedara-Etna* area, *Cefalù-Lascari-Gratteri* area, *Saponara-Gallo* rivers, *Elicona* river, Northern and Southern *Imera* river, *Sosio-Verdura* rivers, *Torto* river and its minor watercourses, *Pollina* river, *Valdemone* area, *Agrò-Savoca-Pagliara* area, *Rosmarino* river, *Castelvetrano* area, Valley of the *Nisi* river, and *Corriolo-Mela* rivers, distributed across all the regional territory. In their present state, strong similarities are apparent as regards the formal contents of each Memorandum of Understanding hitherto undersigned. If this reality may attest to a beneficial exchange of knowledge between the different partnerships, on the one hand, at the same time it might conceal a flawed attention to the unique characteristics of each river basin context, on the other hand, as well as to their distinctive environmental, politico-institutional and socio-economic development contexts.

References

Allain S (2001) Planification participative de basin et gouvernement de l'eau. Géocarrefour 76(3):199–209

Bandinu P, Demuro P, Massa L, Sechi P C (2011) Il processo dell'accordo del Flumini Mannu: un'ipotesi di riqualificazione pesistico-ambientale del contesto agricolo-insediatvo dell'intero bacino. In: Bastiani M (ed) Contratti di Fiume. Pianificazione strategica e partecipata dei bacini idrografici. Approcci, esperienze, casi studio, Dario Flaccovio Editore, Palermo, pp 527–537

Barraqué B (1994) Entre public et privé, l'eau et ses agences. In: Marié M et al (eds) Cultures, usages et stratégies de l'eau en Méditerranée occidentale. L'Harmattan, Paris, pp 371–396

Barraqué B (2003) La direttiva quadro dell'Unione Europea: quali difficoltà di applicazione al Nord e al Sud dell'Europa? In: Ferragina E (ed) Una politica delle risorse idriche per il futuro del Mediterraneo. Il Mulino, Bologna

Bartoli M, Perlini S (2011) Mincio e Oglio: due esperienze parallele a confronto. In: Bastiani M (ed) Contratti di Fiume. Pianificazione strategica e partecipata dei bacini idrografici. Approcci, esperienze, casi studio, Dario Flaccovio Editore, Palermo, pp 238–263

Bastiani M (2011) Dalla valorizzazione degli ambiti fluviali ai contratti di fiume. In: Bastiani M (ed) Contratti di Fiume. Pianificazione strategica e partecipata dei bacini idrografici. Approcci, esperienze, casi studio. Dario Flaccovio Editore, Palermo, pp 3–30

Bianco A, Pineschi G (2011) I contratti di fiume nel contesto normativo europeo e nazionale. In: Bastiani M (ed) Contratti di Fiume. Pianificazione strategica e partecipata dei bacini idrografici. Approcci, esperienze, casi studio, Dario Flaccovio Editore, Palermo, pp. 139–163

Bonamini I, Brugioni M (2011) Piani di distretto e contratti di fiume in Toscana. In: Bastiani M (ed) Contratti di Fiume. Pianificazione strategica e partecipata dei bacini idrografici. Approcci, esperienze, casi studio, Dario Flaccovio Editore, Palermo, pp. 353–362

Bravard J-P (2002) Le «traitement» des versants dans le département de la Drôme. Des inondations de 1840 à la loi du 27 juillet 1860. Annales de Ponts et Chaussées 103:37–43

Briola M (2004) SAGE et contrat de rivière. Courrier des épines drômoise 121

Brun A (2010) Les contrats de rivière en France: un outil de gestion concertée de la ressource en local. In: Schneier-Madanes G (ed) L'eau mondialisée. La gouvernance en question, Éditions La Découverte, Paris, pp 305–321

Brun A (2014) France's water policy: the interest and limits of river contracts. In: Schneier-Madanes G (ed) Globalized water: a question of Governance. Springer, Dordrecht, pp 139–147

Castellana G, Tomaselli G, La Rosa M R (2011) Contratti di fiume nelle aree Parco: verso un contratto di fiume del bacino dell'Alcantara. In: Bastiani M (ed) Contratti di Fiume. Pianificazione strategica e partecipata dei bacini idrografici. Approcci, esperienze, casi studio, Dario Flaccovio Editore, Palermo, pp 505–526

Caudal S, Janin P (2007) La loi du 30 décembre 2006 sur l'eau et les milieux aquatiques entre volontarisme et renoncements. Droit administratif 4

Chémery J-B, Luczyszyn H (2011) Contrats de canaux en Provence Alpes Côte d'Azur. De l'idée à la mise en œuvre, différents enseignements. In Actes du Colloque international LPED/IMEP/GIEST, Université de Provence, Centre Saint Charles, Marseille, France 20–21 janvier 2011

Ciarabelli F(2011) Il Tevere in Alta Umbria: dall'Agenda 21 al contratto di fiume. In: Bastiani M (ed) Contratti di Fiume. Pianificazione strategica e partecipata dei bacini idrografici. Approcci, esperienze, casi studio, Dario Flaccovio Editore, Palermo, pp 383–391

Clarimont S (2009) L'évolution des politiques française et espagnole de l'eau. Entre directive communautaires et décentralisation administrative. Économie rurale 309:34–49

Clemente F, Giannetta A, Porro E (2011) I contratti di fiume e di lago in Piemonte: casi studio ed esperienze. In: Bastiani M (ed) Contratti di Fiume. Pianificazione strategica e partecipata dei bacini idrografici. Approcci, esperienze, casi studio, Dario Flaccovio Editore, Palermo, pp 265–280

Clerici M, Borasio M, Canti F (2011) I contratti di fiume in Lombardia. In: Bastiani M (ed) Contratti di Fiume. Pianificazione strategica e partecipata dei bacini idrografici. Approcci, esperienze, casi studio, Dario Flaccovio Editore, Palermo, pp 219–237

Dervieux A (2005) La difficile gestion globale de l'eau en Camargue (France): le Contrat de delta, VertigO - la revue électronique en sciences de l'environnement [En ligne] 6(3)

Di Federico I (2003) Considerazioni sulla l. 183/1989 ad un decennio dall'emanazione. In: Ferrucci E M (ed) Primo Forum Nazionale. Rischio idraulico e assetto della rete idrografica nella pianificazione di bacino. Questioni, metodi, esperienze a confronto. Maggioli Editore, Rimini, pp 125–146

Drobenko B (2007) La loi sur l'eau et les milieux aquatiques: entre atermoiements et renoncements. Revue juridique de l'environnement 2:141–172

Duport L (1991) De l'aménagement intégré à la gestion globale des rivières: Les contrats de rivière ont 10 ans. La Houille Blanche 7–8:549–554

Ercolini M (2006) Dalle esigenze alle opportunità. La difesa idraulica fluviale occasione per un progetto di "paesaggio terzo". Firenze University Press, Firenze

Ferrucci E M (2003) Primo Forum Nazionale. Rischio idraulico e assetto della rete idrografica nella pianificazione di bacino. Questioni, metodi, esperienze a confronto. Maggioli Editore, Rimini

Filpa A (2009) La pianificazione di bacino, delle aree protette e paesaggistica. In: Filpa A, Talia M (eds) Fondamenti di governo del territorio. Dal piano di tradizione alle nuove pratiche urbanistiche. Carocci Editore, Roma, pp 325–340

Flajolet A (2006) Rapport fait au nom de la commission des affaires économiques, de l'environnement et du territoire sur le projet de loi, adopté par le Sénat, sur l'eau et les milieux aquatiques. Assemblée nationale n. 3070/2005-2006

Gerardi M, Di Fazio G (2011) Programma integrato di interventi per la valorizzazione del bacino idrografico del fiume Noce- recupero e valorizzazione di un territorio ad alta valenza ambientale e turistica. In: Bastiani M (ed) Contratti di Fiume. Pianificazione strategica e partecipata dei bacini idrografici. Approcci, esperienze, casi studio, Dario Flaccovio Editore, Palermo, pp 487–503

Ghiotti S (2006) Les territoires de l'eau et la décentralisation. La gouvernance du bassin versant ou les limites d'une évidence. Développement durable et territoires [En ligne], Dossier 6

Ghiotti S (2010) Le «bon état écologique des eaux» pour 2015: une ambition qui bouscule la gestion des territoires en France. In: Schneier-Madanes G (ed) L'eau mondialisée. La Découverte, Paris, pp 143–159

Goria A, Lugaresi N (2004) The evolution of the water regime in Italy. In: Kissling-Näf I, Kuks S (eds) The evolution of national water regimes in Europe. Kluwer Academic Publishers, pp 265–291

Governa F, Toldo A (2011) Le linee guida dei contratti di fiume in Piemonte. In: Bastiani M (ed) Contratti di Fiume. Pianificazione strategica e partecipata dei bacini idrografici. Approcci, esperienze, casi studio. Dario Flaccovio Editore, Palermo, pp 280–298

Iacoviello M (2001) Pianificazione di Bacino imbrifero e reti di connettività ecologica. In: Rigillo Troncone M (ed), Quaderni di ingegneria ambientale. Inquinamento e depurazione 34

Laganier R, Scarwell H-J, Bartout P, Benmalek Y, Brunaud D, Chow-Toun F, Franchomme M, Parage J, Temam S (2009) Gestion de l'eau, gestion des risques: de récentes recherches pour évaluer la territorialisation de l'action publique. In: Laganier R, Fassetta G A (ed) Les géographies de l'eau. Processus, dynamique et gestion de l'hydrosystème. L'Harmattan, Paris, pp 55–98

Larrue C (2002) La gestion de l'eau à la croisée des politiques publiques et des territoires. Bulletin de l'Association des Géographes Français 1(mars):67–77

Lascoumes P, Le Bourhis J-P (1998) Les politiques de l'eau: enjeux et problématiques. Regards sur l'actualité 241:33–41

Le Bourhis JP (2003) Complexité et trajectoires d'apprentissage dans l'action publique. Revue Internationale de Politique Comparée 10(2):161–175

Magnaghi A (2011) Contratti di fiume e pianificazione: uno strumento innovativo per il governo del territorio. In: Bastiani M (ed) Contratti di Fiume. Pianificazione strategica e partecipata dei bacini idrografici. Approcci, esperienze, casi studio. Dario Flaccovio Editore, Palermo, pp 31–44

Martini E, Di Cesare M, Tortoioli L, (2011) Il progetto Tevere: un esempio di progettualità programmatica tra il PUST e il PPR dell'Umbria. In: Bastiani M (ed) Contratti di Fiume. Pianificazione strategica e partecipata dei bacini idrografici. Approcci, esperienze, casi studio. Dario Flaccovio Editore, Palermo, pp 377–382

Massardier G (2009) La gouvernance de l'eau: entre procédure de concertation et régulation «adhocratique». Le cas de la gestion de la rivière Verdon en France. VertigO - la revue électronique en sciences de l'environnement [En ligne], Hors série 6

Menduni G (2007) La pianificazione a scala di bacino tra governo delle risorse e governo delle trasformazioni. In: Ercolini M (ed) Fiume, paesaggio, difesa del suolo. Superare le emergenze, cogliere le opportunità. Atti del Convegno Internazionale Firenze, 10-11 maggio 2006. Firenze University Press, Firenze, pp 93-100

Montaletti V (2011) Processi integrati di valorizzazione fluviale nella programmazione regionale. In: Bastiani M (ed) Contratti di Fiume. Pianificazione strategica e partecipata dei bacini idrografici. Approcci, esperienze, casi studio. Dario Flaccovio Editore, Palermo, pp 315–323

Nicolazo JL, Redaud JL (2007) Les agences de l'eau: Quarante ans de politique de l'eau. Editions Johanet, Paris

Pezon C (2006) The French PPP model for water service management: genesis and key factors of success. Urban Water Conflict, International Hydrological Program (IHP), UNESCO, Working Series, SC-2006/WS/19, Paris

Pizziolo G, Micarelli R (2011) Il contratto di fiume/paesaggio del medio Panaro. In: Bastiani M (ed) Contratti di Fiume. Pianificazione strategica e partecipata dei bacini idrografici. Approcci, esperienze, casi studio. Dario Flaccovio Editore, Palermo, pp 323–343

Puech D, Boisson J-M (1995) Eau-ressource et eau-milieu. Une interdépendance croissante impliquant une évolution des modalités de gestion. Eau-ressource et eau-milieu vers une gestion durable, Les cahiers de l'Economie Méridionale, coll. Rapports d'étude 1:5–47

Richard A (2005) Vers l'institutionnalisation d'une gouvernance de bassin? Le cas du bassin versant de l'Orb. In: Le Goulven P, Bouarfa S, Kuper M, (eds) Gestion intégrée de l'eau au sein d'un bassin versant. Actes de l'atelier du PCSI, 2-3 décembre 2003, Montpellier, France

Scarwell H-J (2007) Espaces et échelles d'intervention: la gestion de l'eau du XVIIIe siècle à nos jours dans le Nord Pas-de Calais. In: Scarwell H-J, Kergomard C, Laganier R (eds) Environnement et gouvernance des territoires – enjeux, expériences et perspectives en région Nord-Pas de Calais. Presses Uviversitaires du Septentrion, Villeneuve d'Ascq (France), pp 39–87

Scarwell H J, Laganier R (2004) Les recompositions de l'action publique en contexte transfrontalier: l'exemple du district hydrographique de l'Escaut. Revue Développement Durable et Territoires, dossier 2 Gouvernance locale et Développement durable

Urbani P (2007) Acque, fiumi e difesa del suolo: problemi giuridico-legislativi degli assetti della pianificazione e della tutela ambientale. In: Ercolini M (ed) Fiume, paesaggio, difesa del suolo. Superare le emergenze, cogliere le opportunità. Atti del Convegno Internazionale Firenze, 10-11 maggio 2006, Firenze University Press, Firenze, pp 77-83

Venti D, La Pegna M P (2011) Verso un contratto di fiume per il fiume Nera. In: Bastiani M (ed) Contratti di Fiume. Pianificazione strategica e partecipata dei bacini idrografici. Approcci, esperienze, casi studio. Dario Flaccovio Editore, Palermo, pp 391–399

Zazzi M (2009) Prospettive per il governo dei bacini idrografici. ASUR 96:71–91

Chapter 4
Case Studies

Abstract This chapter illustrates precisely the comparison between the two pairs of case studies, highlighting differences and analogies between France and Italy in terms of evolution, dissemination and application of the river contract model. In order to describe and analyze some paradigmatic examples both concerning hydrographic basins located in metropolitan areas and other ones characterized by a largely rural aspects, four case studies have been specifically selected among river contracts implemented in the two national contexts: the *Contrat de Rivière Yzeron* (Rhône-Alpes) and the *Olona-Bozzente-Lura River Contract* (Lombardia), on the one hand, and the *Contrat de bassin de la Basse Vallée de l'Ain* (Rhône-Alpes) and the *Val d'Ofanto River Contract* (Puglia), on the other.

4.1 Introduction

A comparative analysis of the experiences undertaken in France and Italy was the basis for selecting a number of case studies deemed particularly significant for the purpose of verifying the theoretical hypotheses of this research endeavor.

For France, the choice fell on two *contrats de rivière* (*CdR*) stipulated for the *Région Rhône-Alpes* that also happens to comprise 48 % of all French experiences. The first case study selected was that of the *Contrat de rivière Yzeron*, concerning the homonymous river basin located in the north-western part of the region, just west of the city of Lyon. Specifically, it offers an example of a *CdR* implemented within a metropolitan territorial context, moreover, in the absence of a *Schéma d'Aménagement et Gestion des Eaux* (*SAGE*) already in force. The second case examined was that of the *Contrat de bassin de la Basse Vallée de l'Ain* relative to an area with a strong rural connotation, located in the northern part of the *Région Rhône-Alpes*, north-east of the city of Lyon. In the literature, this case study is cited as a best practice in light of the elevated levels of integration achieved, in this

M.L. Scaduto, *River Contracts and Integrated Water Management in Europe*,
UNIPA Springer Series, DOI 10.1007/978-3-319-42628-0_4

territorial context, between the instruments of water resources management and those of urban and territorial planning.

Even with regard to Italy, two case studies were opportunely selected with respect to two particularly meaningful basin contexts, namely the hydrographic system of the sub-basins of the *Olona*, the *Bozzente* and the *Lura* rivers, and the basin of the *Ofanto* river. By and large, the former case historically represents the first river contract (RC) experience in Italy and refers to a sub-basin of the Po river located in the north-western part of the *Regione Lombardia*, north of city of Milan.

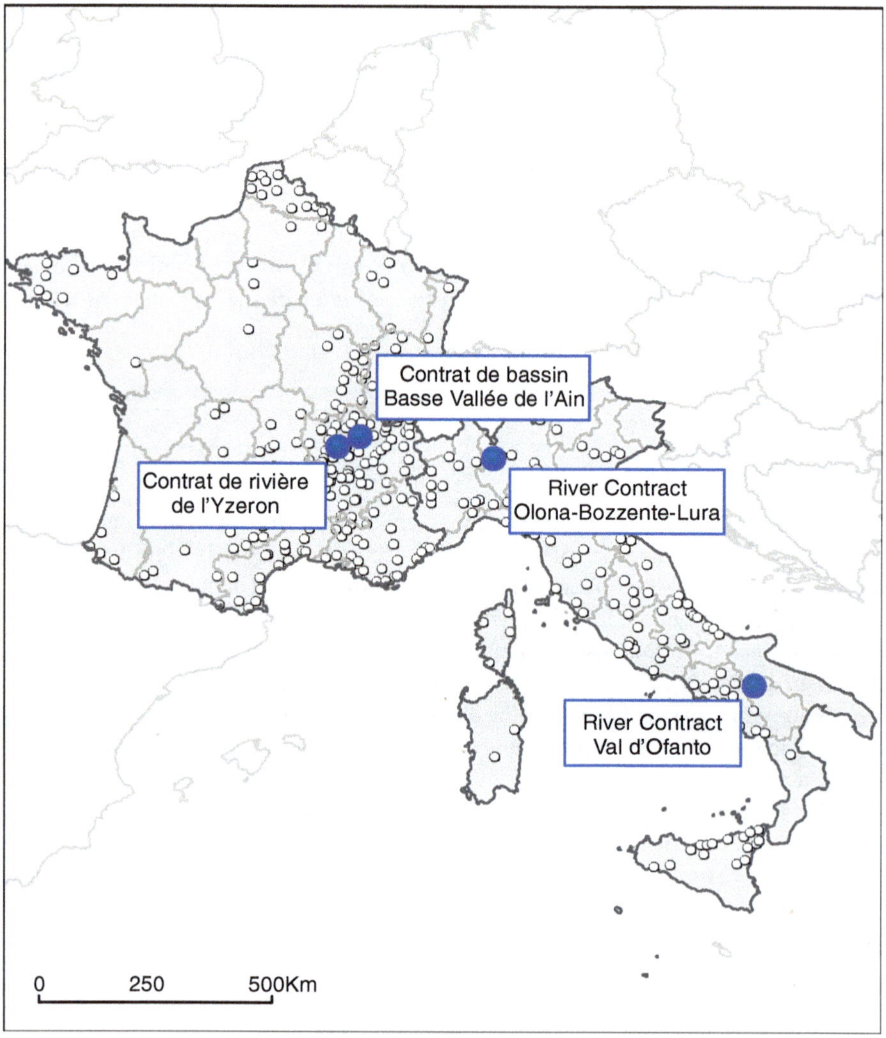

Fig. 4.1 The selected case studies in France and Italy

On the other hand, the latter case study, albeit still in the making, refers to the *Val d'Ofanto River Contract*, which impacts the vast geographical area comprised between *Campania*, *Basilicata* and *Puglia*, in southern Italy.

On further analysis of these four selected case studies (Fig. 4.1), drawing comparisons was possible on several levels, first of all between RC concerning hydrographic basins located in metropolitan areas in contrast to those characterized by a largely rural aspects. Secondly, the comparative analysis permitted an initial framework of comparison between the two national contexts of France and Italy, in terms of evolution, dissemination and application of the RC model, especially with reference to the respective policies and forms of territorial governance from the standpoint of integrated water resources management. Specifically, this evaluation is based on a pairwise comparison, on the one hand contrasting the *Contrat de Rivière Yzeron* with the *Olona-Bozzente-Lura River Contract*, while on the other comparing the *Contrat de bassin de la Basse Vallée de l'Ain* to the *Val d'Ofanto River Contract*.

4.2 River Contracts in Urbanized Contexts: The *Yzeron* and the *Olona-Bozzente-Lura* Case Studies

The two case studies regarding metropolitan contexts refer to the *Yzeron* river basin, right bank tributary of the *Rhône*, and the *Olona* sub-basin and its tributaries *Bozzente-Lura*, located in the central-northern part of the largest hydrographic basin of the Po river. In either case the sub-basins feature urban settlement patterns concerning large metropolitan areas (with over 33 % urbanized areas), i.e. Lyon and Milan (population: 1,200,000 and 1,300,000, respectively), comprised in the hydrographic territories of the *Rhône* and the *Po*, equally having considerable importance and size (95,000 and 71,000 km^2, respectively).

As a result of the relationships with their respective metropolitan areas, the increasing anthropic pressures due to development and urbanization, as well as the often misguided practices of land taken and soil sealing, have in either case determined a significant deterioration in the quality of water resources and an increase in flood hazards, including inundations and erosion of river beds and banks. In addition to the latter, it must be also considered the impact of prolonged periods of low water levels and the phenomena of diminishing biodiversity and degradation of river environments, even if there is a persistence of a number of areas of great ecological and landscape interest. The latter have not only increased concern about these environmental issues, but also led to efforts to protect and preserve the natural integrity of these environments (Piegay et al. 2000; Lafont 2006; Regione Lombardia-IREALP 2010).

4.2.1 *The* Contrat de Rivière de l'Yzeron

The *Yzeron* basin, with its 150 km^2 extending over the entire territory of 26 municipalities, is considered a prime example of small peri-urban basin (Breil et al. 2006; Lafont 2006; Breil 2007) (Fig. 4.2). As of the twentieth century, this territory has been directly subjected to urbanization measures underlying the expansion of

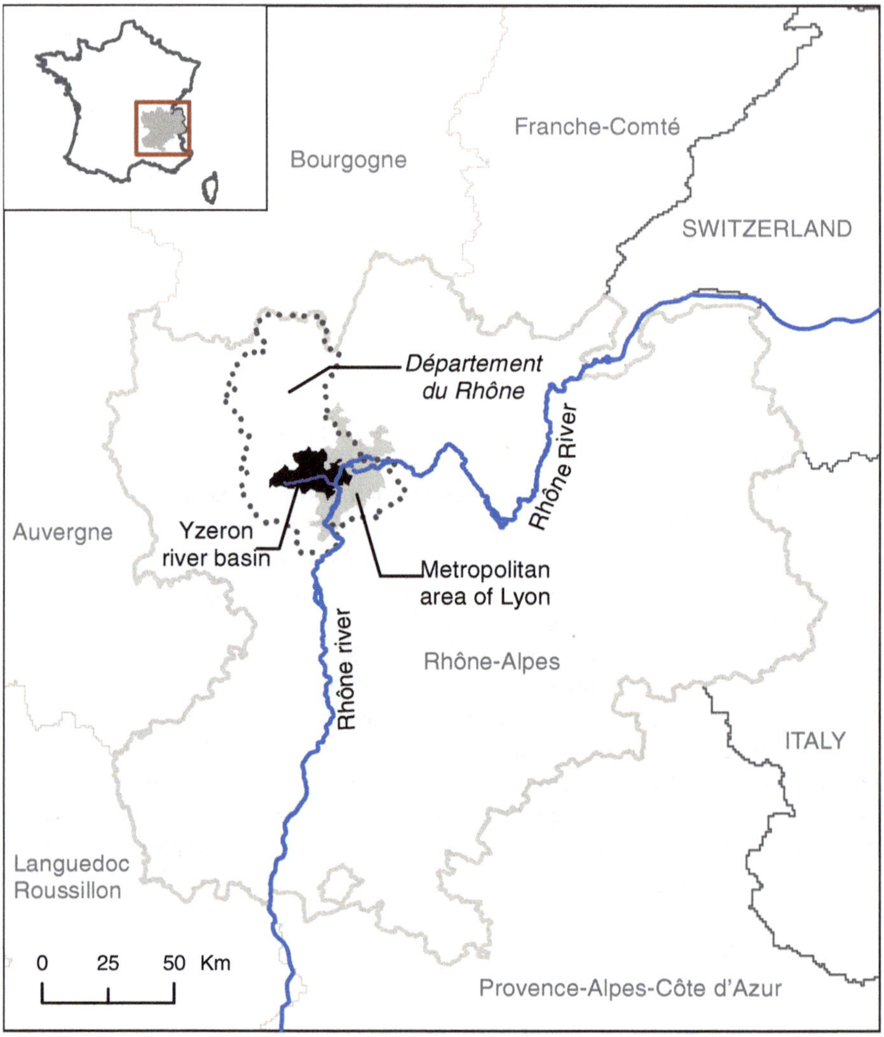

Fig. 4.2 The Yzeron basin in the regional context of Rhône-Alpes

Lyon, particularly of its western suburbs, determining extensive soil sealing as well as the abandonment of agricultural areas (Radojevic et al. 2002; Thollet and Branger 2009).

In the *Yzeron* basin the urbanization gradient increases as one proceeds downstream. The mountainous part, represented by the countryside of the *Monts du Lyonnais*, is characterized by small villages with a strong rural connotation. The intermediate zone midway between the mountains and the valley bottom comprises the *Ouest Lyonnais*, on the residential outskirts of the city, characterized by peri-urban landscapes (Fig. 4.3). Finally, the basin valley spans until the confluence with the *Rhône* river, and is fully encompassed within the city limits of Lyon. Consequently, in these urbanized areas, the course of the *Yzeron* has undergone profound mutations, resulting laden with artifacts to some extent, even to the point that its bed and banks have often been concreted over (Fig. 4.4).

As far back as the 1980s, several studies conducted by the *Groupement Rhône-Alpes Infrastructure et Eau* had already reported on the principal damaging effects determined by the high levels of urbanization of this area (Lalo 1986; Hubert 1988; Meuret 1988).

Following the floods of 1983, 1989, 1990 and especially of 1993, the need to step in with a coherent and global approach for solving management problems at the river basin scale became generally acknowledged among *riverains*, local authorities and institutional representatives alike. The offshoot of this strengthened resolve, regarding water resources management at the basin scale, was the establishment in 1991 of the *Syndicat d'Etude pour l'Aménagement et la Gestion de*

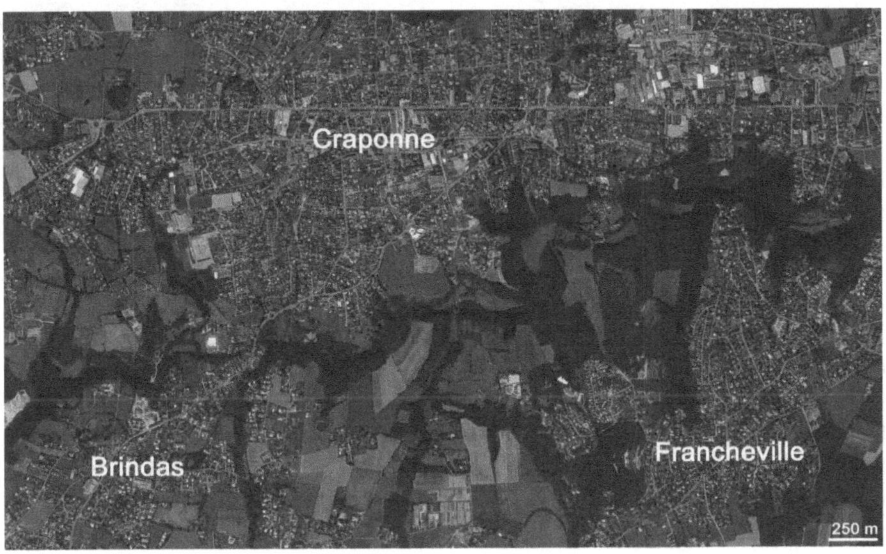

Fig. 4.3 The Yzeron river flowing through peri-urban landscape, characterizing the zone between mountains and the bottom valley (*basemap* Microsoft Bing)

Fig. 4.4 The Yzeron river at the confluence with the Rhône river, in the southern urbanized part of Lyon (*basemap* Microsoft Bing)

l'Yzeron, du Ratier et du Charbonnières (*SEAGYRC*). This intercommunal-institution, formed by 12 municipalities of the valley basin area, was set up with the precise task of defining priorities for the protection against hydrogeological risks as well as for of the integrated management of common water resources. In order to identify a strategy capable of achieving a broad consensus among all stakeholders, the *SEAGYRC* conducted a study in 1993 entitled and aimed at the *Définition d'une stratégie d'aménagement sur le périmètre du bassin versant de l'Yzeron*. The results of the investigation highlighted the importance of up-stream-down-stream interconnections for the management of watercourses, so as to ensure greater coherence amongst actions carried out in various areas of the basin. The ultimate goal of this concerted strategy was clearly to minimize the risk of hydrogeological hazards, especially for the inhabitants of the valley, by overcoming the perspective of action programs often rigidly conditioned by administrative boundaries (SAGYRC 2007).

The *Contrat de Rivière de l'Yzeron* was about 20 years in the making, from the institution of the *SEAGYRC* in 1991–2011, the year the *Étude de bilan, évaluation et prospective* was completed. The breakdown of the timeline, until the signing of the agreement in 2002, highlights to what extent the consultation phase, at the basis of the activation of the *CdR*, was particularly time-consuming and laborious. This lengthy initial phase, which lasted more than ten years, reflects the many stumbling blocks encountered throughout in order to reconcile the different views of the up-stream and down-stream basin communities, particularly with regard to the risk of flooding (GRAIE-ZABR 2008). In fact, the complex and differentiated make-up of the *Yzeron* river basin is the underlying reason for the numerous conflicts between the various territorial contexts as well as between the up-stream and down-stream communities (Radojevic et al. 2002; SAGYRC 2008).

The very nature of this particular territorial and administrative context precisely required the involvement, already in the activation stages of the *CdR*, of a host of diversified public and private entities. Among these, the leading role was assumed by the *Syndicat d'Aménagement et Gestion de l'Yzeron, du Ratier et du Charbonnieres* (*SAGYRC*), established in 2001 based on the previous *SEAGYRC* founded in 1991. This organization, also known as the *Syndicat intercommunal du bassin de l'Yzeron*, consists of representatives from most of the 26 municipalities of the entire hydrographic unit. It still constitutes the permanent administrative body of water resources management at the basin scale and, for this reason, has been vested with the competences not only regarding the actualization of the *Contrat de Rivière de l'Yzeron*, but also in all matters pertaining to the protection, quality-status improvement and management of river environments.

In the course of the actualization of the *CdR*, the *SAGYRC* assumed the role of *structure porteuse*, coordinating the activities of the *Comité de rivière* and representing the *maître d'ouvrage* (project manager) of the action program.

Even the State, the *Agence de l'eau*, the *Région Rhône-Alpes*, the *Département du Rhône*, the Metropolitan City of Lyon, also known as *Grand Lyon*, the intercommunal structures within the *Yzeron* basin and municipalities have contributed substantially to the implementation of the *CdR*. In particular, all these subjects have participated to varying extents in financing the program of action.

Other territorial interest groups, such as associations of farmers, fishermen and of the industrial sector as well as environmentalist associations, also took active part in the *Comité de rivière* and its thematic sub-committees (SAGYRC 2008).

Another aspect worth highlighting is the fact that the *Yzeron* hydrographic basin is also subject to the administrative and management competences of various intercommunal bodies and structures. Besides the *SAGYRC*, the presence of eight intercommunal structures, which in some cases have taken on the role of *maitres d'ouvrage* of some of the actions of the contract, clearly shows how the *Contrat de Rivière de l'Yzeron* represents a milestone in terms of territorial consultation, institutional coordination and the building of an effective *solidarité amont-aval*, in spite of the inevitable inter-institutional pitfalls. To this effect, a case in point is the institutional relations existing between *SAGYRC* and *Grand Lyon* during the implementation of the action program. In fact, despite *Grand Lyon* being heavily involved in the *CdR*, by way of a technical and financial partnership with the *SAGYRC* and in particular regards the protection from the risk of flooding of built-up areas in the valley area, the political relations and the technical collaboration between the two entities proved quite complicated, among other reasons, for the lack of any political representation of *Grand Lyon* within the *Comité de rivière*.

The protracted elaboration procedure of the *Contrat de Rivière de l'Yzeron* was divided into five phases: (I) the formation and the establishment in 1998 of the *Comité de rivière*, following the presentation in 1997 of the *Etude préalable d'aménagement* by the *SEAGYRC* and the subsequent approval of the *Contrat de Rivière de l'Yzeron*; (II) the drafting between 1999 and 2000 of detailed studies focusing on landscape, quality status and use of water resources, waste to water management practices, analyses of the sources of agricultural pollution and management of resources for fisheries, so as to be able to elucidate the objectives and program of actions in detail; (III) the institution in 2001 of the *SAGYRC*, in charge of drawing up the final dossier of the RC; (IV) implementation of the action plan, starting from the signing of the *CdR* in 2002 until its completion in 2008; (V) reporting on overall conclusions and evaluation in terms of accomplishments (2009–2011).

The evaluation procedure conducted in this latter phase provided opportunities to examine with hindsight the results regarding the implementation of the *Contrat de Rivière de l'Yzeron*. Specifically, the original program included 115 actions, funded with a total investment of 41.3 million euro, articulated into four thematic sections, so called *volets*: *Volet A*, focusing on improving the quality of surface waters; *Volet B1*, geared to restoring the natural equilibrium and hydrological regimes, also by way of actions striving to improve the accessibility to the river; *Volet B2*, dedicated to mitigating the risk of flooding; *Volet C*, focusing on territorial consultation and on actions to inform local communities and raise public awareness, such that the environmental requalification initiatives and these new forms of river basin integrated management may effectively afford long-lasting effects.

However, a fraction of the planned actions never actually came to fruition for various reasons based in part on shortcomings in the economic and technical forecasts, which proved inadequate for the required interventions. In addition, the final results of the initiative were partially undermined by other factors related, on the one hand, to the effects of flooding occurring during the implementation phase of the program of action and, on the other hand, to unforeseen alters in the institutional profile of the *SAGYRC*, which entailed surrendering some of its authority in favor of other bodies. Last but not least, some opposition to the interventions even came from local players, representing a further drag on the overall outcomes of the *CdR* (BURGEAP 2011).

Altogether these reasons served to highlight the drawbacks of allotting a mere five-year timeframe for programming, which proved the least bit compatible with natural river dynamics or the cultural and socio-economic processes in progress in the territories of the *Yzeron* river basin. Indeed, the issue of protection from floods and mitigating the risk of inundation to the area was seen as the overriding priority of the program of actions, thus significantly skewing results to the point of failing to ensure the proper allocation of financial resources or achieve the overall objectives.

From this perspective, it can be affirmed that an interim evaluation conducted midway through the project, would have allowed for readjustments to the action plan, supposedly enabling participants to more effectively realign the priorities of the planned actions so as to maximize outcomes.

In 2008, upon concluding the *CdR*, political representatives of the local authorities opted to refrain from promoting a second contract or embarking upon the procedure for drawing up a *Schéma d'Aménagement et Gestion des Eaux*, in light of the conclusions of the study of 2011 detailing the overall evaluation in terms of accomplishments, but also due to the impending political elections of 2014. At present, the interventions initiated, then left uncompleted within the life span of the *CdR*, in particular those concerning *Volet B2* in matters of flood protection, are once again underway thanks to funding expressly granted to the *SAGYRC* by the *Région Rhône Alpes*, by way of an ad hoc institutional agreement.

Although the implementation of the *Contrat de Rivière de l'Yzeron* undeniably had its share of shortcomings, the conclusions of the overall evaluation in terms of accomplishments did, however, credit it with having occasioned a useful territorial consultation process. Moreover, it highlighted its aptitude for increasing the awareness of the need to systematize an approach able to reconcile the mountain-community demands with those of their valley counterparts, out of a sense of true *solidarité amont-aval* (BURGEAP 2011). From this viewpoint, the *Contrat de Rivière de l'Yzeron* constitutes a noteworthy political, administrative and management experiment that has contributed to overcoming the constraints of programming and planning initiatives, still overly hampered by municipal administrative confines. Such promise is arguably one of the greatest strengths of the *CdR*, with a potential for greater consistency in matters of urban and territorial development, while protecting and enhancing the quality standards of fluvial environments, in general, within the territories concerned.

It is also worth underscoring the fact that the 2011 analysis of the overall conclusions and the evaluation of action program undertaken, reasserted the importance of the river-basin paradigm for urban and territorial planning, and integrated water management, as well.

Even for this reason, it is interesting to take into consideration the horizontal and vertical relations established between the *Contrat de Rivière de l'Yzeron* and other existing planning tools regarding the inter-municipal level. In fact, due to their mutual relation with the territorial scale of the river basin, they may, over time, determine particularly significant interdependencies. In particular, it seemed warranted to delve into the relations between the *CdR* and the *Schémas de Cohérence Territorial* (*SCOT*). The latter instruments, adopting the guidelines set by the *SDAGE* and *SAGE*, represent the superordinate tools by which priorities for the integrated management of water resources and related ecosystems can be transferred to the urban and territorial planning at the municipal scale.

The *Yzeron* river basin is under the influence of diverse instruments that, for various reasons and to varying extents, concern its hydrographic and environmental components as well as its local socio-economic systems.

The *Directive Territoriale d'Aménagement* (*DTA*) *de l'Aire Metropolitan Lyonnaise*, approved in 2007, impacts the vast territory of the three urban agglomerations of Lyon, *Saint-Etienne* and *Nord-Iseère*, the smaller urban agglomerations of *Villefranche-sur-Saone*, *Givors*, *Vienne*, *Ambérieu* and *Pont-de-Chéruy* and numerous rural centers, in addition to areas of agricultural and naturalistic value. With regard to water resources, the primary objective of the *DTA* is the well-balanced and concerted management of surface and underground waters, yet mindful of their different uses. Furthermore, considerable attention is paid to environmental rehabilitation, with particular regard to the impact of urban run-off, industrial and agricultural wastewater discharges, as well as to contrasting the risk of floods. In relation to the latter, the *DTA* has underscored the need to safeguard flood-prone areas, also by mapping out river corridors.

It is precisely with respect to these specific objectives of the *DTA* that major links with the *Contrat de Rivière de l'Yzeron* can be established, in particular regarding the above mentioned *Volet B2*, focused on protection from the hydrogeological risk. To this effect, of particular significance is the fact that the *DTA* has highlighted the need to encompass the *Plans Prévention des Risques Inondation* (*PPRI*) into an integrated flood prevention policy, on the one hand modeled on the basin scale, and on the other hand based on stronger correlations with the *SDAGE*, *SAGE* and *CdR*.

Even in the case of the *Schéma de Cohérence Territoriale* (*SCOT*) *de l'Agglomération Lyonnaise*, approved in 2011, the prevailing mutual aim with the *Contrat de Rivière de l'Yzeron* remains the mitigation of the risk of floods and landslides. In fact, in line with the objectives detailed in the *SDAGE* and *PPRI*, this *SCOT* advances orientations on the matter with the aim of safeguarding the flood-prone areas within the limits of a number of municipalities, comprised within *Grand Lyon*, *Est Lyonnais* and *Val d'Ozon*. In this sense, in the *SCOT*, the opportunity to forge solid bonds with *SAGE* and *CdR* was highlighted, both among the programs of actions envisioned and in terms of creating synergies among the respective active counterparts.

With its twenty-year time-frame perspective, the *SCOT* acknowledges the significant role of waterways in structuring the territory of the metropolitan area, underlining its landscape, ecological, socio-economic and recreational values. In this context, the rivers become the structural constituents of the so called *réseau bleu*, playing a vital role in the intricacies of the ecosystem whilst replenishing the water supply of the *Grand Lyon* urban agglomeration.

The reference area of the *Contrat de Rivière de l'Yzeron* is also subject to the *SCOT de l'Ouest Lyonnais*, whose territorial scope comprises the *Pays de l'Arbresle*, *Vallons du Lyonnais*, *Pays Mornantais* and the *Vallée du Garon*, immediately west of the *Grand Lyon* metropolitan area.

In the *Rapport de présentation* of this *SCOT*, the thematic section on the quality and management of water resources is outstanding and makes ample direct references to the contents of the European Water Framework Directive and the *SDAGE Rhône-Méditerranée*. Specifically, the *SCOT* fosters safeguarding the so called *trame verte et bleue*, but also highlights the crucial role, in attaining the objectives of integrated water resources management policies, of the five *Contrats de rivère Yzeron*, *Azergeues*, *Brévenne-Turdine*, *Garon*, and *Gier*, which overlap with the reference area of the SCOT.

A further supra-municipal tool of urban and territorial planning affecting the *Yzeron* river basin is represented by the *Plan Local d'Urbanisme* (*PLU*) *du Grand Lyon*. It defines the orientations for sustainable development, so as to be integrated and coherent with the territory of 58 municipalities that compose the metropolitan agglomeration. The *PLU* identifies three major orientations which include the need to (I) promote the development of the metropolitan area while fully respecting the natural environment, (II) strengthen social cohesion and *mixité*, and (III) foster the development of local economic systems. These general objectives constitute unifying elements with the *Contrat de Rivière de l'Yzeron*, especially with regard to safeguarding water quality from pollutant-laden discharges, protection from natural hazards, as well as pursuing more befitting planning patterns for new urban development projects.

With regard to the basin planning tools in use throughout the territory of the *Yzeron*, the *Schéma Directeur d'Aménagement et Gestion des Eaux* (*SDAGE*) *Bassin Rhône-Méditerranée 2016–2021* reaffirms the main direction for protecting and enhancing the quality standards of aquatic environments at the basin scale, as stated in the previous *SDAGE 2010–2015*. As its foremost priority, the latter plan had put forth achieving a good water quality status by 2015, in compliance with the dictates of the *European Water Framework Directive*.

It is important to note that the orientations of the *SDAGE* have prescriptive value with reference to the provisions concerning water resources included in the *SAGE*, *SCOT* and *PLU*. In particular, the *SDAGE* tends to foster local management of water resources and to ensure coherence between urban and territorial planning and water resources management, stressing the need to ensure and bolster coherence among planning tools, whilst striving to build engaging relationships between those institutional actors directly responsible and all other territorial stakeholders.

In the thematic section of *SDAGE 2010–2015*, dedicated to identifying specific measures for each territory, the main affinities between the *SDAGE* and the *Contrat de Rivière de l'Yzeron* emerge with specific regard to the containing agricultural sources of pollution, stepping up up-stream and down-stream interconnections, curbing patterns of water overconsumption and managing hydraulic works.

4.2.2 *The* Olona-Bozzente-Lura *River Contract*

The *Olona* river flows in the eastern section of the metropolitan area of Milan, in the vicinity of its most urbanized and industrialized districts (Fig. 4.5). Its basin extends for 370 km^2 in the transition zone between the mountainous-hilly sectors of the Provinces of *Varese* and *Como*, and the high plains of the Province of Milan (Calori 2004; Regione Lombardia-ARPA Lombardia 2004). The hydrographic system, consisting in the *Olona* river and its two left-bank tributaries, *Bozzente* and *Lura*, is characterized by difficult interactions between urbanized and natural systems, against a backdrop of high levels of urbanization intensity (Ferraresi and Magnaghi 1992; Magnaghi 1995, 2004) (Fig. 4.6). The latter process began in the *Olona* valley in the early ninetieth century as part of the territorial changes that, on a larger scale, manifested in the river basins of the *Olona* and *Lambro* with increasing phenomena of human settlement and industrial development driven to the river banks, impacting heavily on the risk of flooding, on water quality as well as on the surrounding environment. As a result of this process, morphological alterations of watercourses ensued. In the case of the *Olona*, which previously flowed directly into the *Po* river, its course was diverted to the city of Milan and its waters henceforth have flowed into the *Lambro*.

Fig. 4.5 The Olona-Bozzente-Lura basin in the regional context of Lombardia

From a geomorphological point of view they can be distinguished two areas of the *Olona* river, one consisting of the mountainside headwaters of the basin, up to the town of *Ponte Gurone*, and the other at the valley bottom which leads to the city of Milan. On the mountainside area a greater degree of urbanization is present to the west, represented by the towns of *Varese* and *Induno Olona*, while to the east the territory is mostly made up of farming and forest areas. In correspondence of the *Olona* Valley, there is an alternation between more or less intensively urbanized areas, along with industrial zones located in close proximity to the river.

Fig. 4.6 The Olona river flowing through the north-western part of the metropolitan area of Milan, within the municipality of Rho (*basemap* Microsoft Bing)

The territory of the *Olona* river basin is characterized by extensive fragmentation of its peri-urban agricultural areas and by degradation of its rural landscapes. In addition, the development therein of inhabited settlements and zones of industrial activity, in the absence of proper planning, together with the river engineering interventions performed on its bed, led to its inclusion among the *Area at high risk of environmental crisis* on the part of the Ministry of Environment, under the Law No. 349/1986 (Ferraresi and Magnaghi 1992). However, this, all but enviable, ranking triggered a complex undertaking to research and investigate the matter, culminating in the drafting of a *Piano quinquennale di disinquinamento del bacino idrografico dei fiumi Lambro, Seveso e Olona* i.e. a five-year plan to de-pollute the basinis of the *Lambro*, *Seveso* and *Olona* rivers, authored by the Ministry of Environment in 1988.

Nevertheless, upon independently evaluating the aforementioned emergency plan, in 1993 the *Regione Lombardia* commissioned its *Regional Research Institute* to supplement the plan with their own investigations aimed at identifying the causal factors of environmental degradation and to draw up a proposal of more appropriate solutions (Magnaghi 1995, 2004; Calori 2004). These studies highlighted the purported inadequacy and merely emergency character of the five-year plan of 1988. In this sense, these studies represented but the *incipit* of a lengthy investigative process promoted by the *Regione Lombardia*. Ultimately, upon completion of further analyses, multi-disciplinary and multi-sectoral research, it culminated in the drafting in 2002 of the Memorandum of Understanding entitled *Onwards to River Contracts*.

The two parallel initiatives, activated on the one hand by the Ministry of Environment while on the other by the Regional administration, laid the foundations of the path, from the definition up to the signing, conducive to the *Olona-Bozzente-Lura River Contract*, thus activating in 2004 the first Italian RC experience, whose implementation is still underway.

Under Regional Law No. 26/2003 on *Disciplina dei servizi locali di interesse economico generale. Norme in materia di gestione dei rifiuti, di energia, di utilizzo del sottosuolo e di risorse idriche*, i.e. regulations on local services of general economic interest and on waste and energy management, use of subsoil and water resources, the *Olona-Bozzente-Lura River Contract* was included among the negotiated programming instruments that contribute to the protection and enhancement of water resources and aquatic environments as well as flood protection, promoting coordination and integration among the different policies at the basin and sub-basin scale and the active participation of public and private entities.

Ever since its initial stages, beginning with the Memorandum of Understanding of 2002, continuing throughout a specific research program and the negotiation initiative entitled *River Contracts*, both launched in 2003, the *Regione Lombardia* had played a leading role in the process that led to the signing in 2004 of the *Olona-Bozzente-Lura River Contract*. In particular, the *Regional General Directorate of Public Utilities* oversaw the preliminary information and communication activities and promoted the building of the partnership (Clerici et al. 2011). In this regard, the Regional Administration assumed the tasks of coordinating subscribers, monitoring the implementation of the measures, according to the terms of the contract, while drawing up and dispatching semi-annual progress reports on behalf of the coordination committee. The latter, whose composition includes mayors and representatives of its subscribing local public administration, still plays a key organizational role. Furthermore, it is vested with the authority to approve the program of actions, monitor the implementation phases, make any alters to and update the action plan. In addition, it fosters the involvement of interested public bodies, while facilitating their participation, but also organizes occasions for discussion and information open to various public and private actors throughout the territory (Regione Lombardia 2004).

The Regional Administration and the coordination committee are assisted by a technical committee which, availing itself of the technical, scientific and organizational support of the Regional Environmental Protection Agency (ARPA) of Lombardia, and the Po River Basin Authority, has taken on the task of setting up specific working groups regarding individual issues addressed in the contract. The other parties involved in the *Olona-Bozzente-Lura River Contract*, also as funding agencies, are the municipalities and provinces of the territory, and the so called *Ambiti Territoriali Ottimali* (ATO) of *Milan*, *Varese* and *Como*, the *ARPA* of *Lombardia*, the Basin Authority of the Po river, the *Interregional Agency for the Po river*. In addition, as part of the initiatives to raise awareness in the schools in the territory, the *Regional School Office* was also involved.

A major role is also played by Regional parks and in particular the *Parchi Locali di Interesse Sovra-comunale* (*PLIS in the Italian acronym*). The latter, arising from agreements between municipalities, have proved to be of considerable strategic value in the environmental protection and requalification policies of the territory.

Although several measures to promote new synergies between all public and private territorial players have been deployed, in terms of financial resources, participation on the part of the private sector has remained quite limited. The underlying reason likely consists in the predominant role of the broader and more complex public partnerships, but also in the sectoral perspective of private actors that basically limits the scope of the private funding to the scale of individual landholdings.

The elaboration and activation of the *Olona-Bozzente-Lura River Contract* was structured into six phases: (I) mapping out the preliminary knowledge base; (II) defining a descriptive and interpretative strategic scenario for the medium and long terms; (III) developing and applying an evaluation model for the policies either in effect or being planned; (IV) sketching out a program of actions to carry out the strategic scenario; (V) laying down and implementing a plan of communication, training and education measures; (VI) monitoring the implementation of the program.

Accordingly, during the first phase, the knowledge framework was mapped out focusing on the territory's resources and critical issues, local policy trends and projects already underway. The conclusions of this preliminary study were compiled into a descriptive atlas of sub-basins that served as the groundwork on which to base the ensuing strategies for the medium and long terms. Indeed, from the drafting of this atlas derived a systematic and comprehensive overview of the existing knowledge, even of the policies already in place, so as to identify priority areas and the objectives to be targeted (Regione Lombardia 2004; Clerici et al. 2011).

During the second phase, which aimed at defining the strategic scenario, appropriate structures for governance of the agreement were identified, with reference to the three sub-basins. Additionally, the need for effective interventions for the requalification of landscapes and environments had to be taken into account, while at the same time avoiding, to the extent possible, any overlap of institutional competences on individual issues. To this end, the *PLIS* were identified as the authorities earmarked for supporting these synergistic actions among municipalities, particularly by mapping out multi-purpose river corridors via participated planning initiatives, so as to set up specific programs for the concerned river areas and their surrounding territories.

After developing and applying the evaluation frame of reference to the policies in force, or still in planning, the activation process of the *Olona-Bozzente-Lura River Contract* entered its next phase, during which the program of actions was set down, in light of the strategic scenario outlined.

According to the framework of measures adopted by the *River Po District Management Plan*, four strategic objectives were identified: (I) reduction of water pollution; (II) mitigation of flood risk; (III) requalification of environmental, landscape and settlement systems, regarding the *multi-functional river corridors* of the *Olona*, the *Bozzente* and the *Lura*; (IV) information sharing and dissemination of a new *water culture*.

In particular, the third objective constitutes a priority plank in the program of actions, given that multi-functional river corridors represent load-bearing girders in the fluvial system, also considering the two new territorial hubs, namely the new Trade Fair of Milan and Expo 2015 site (Regione Lombardia-IREALP 2010).

The first program of actions, initiated in 2004 and concluded in 2009, represented the first phase in the implementation process of the strategic scenario outlined in the *Olona-Bozzente-Lura River Contract*. The program focused on measures to safeguard water resources, control emission levels of pollutants, manage flood risk, assure the operational adequacy of bridges and hydraulic works, promote sustainable development of the territory, promote regional parks and *PLIS*, also through initiatives aimed at enhancing environmental quality standards, information campaigns and involvement of stakeholders and local communities (Regione Lombardia 2004). The implementation of the first program of actions met some impediments related primarily to the sheer territorial vastness of the *Olona-Bozzente-Lura* system, the huge number of institutions involved and, therefore, the periodic alternation of parties to, and political representatives in the partnership.

In 2014, upon completing a second season of interventions, the Regional Administration approved the third program of actions regarding the *Olona-Bozzente-Lura River Contract*, allocating approximately 330 million euro.

For all planned actions a path of monitoring activities was also foreseen as well as programmed encounters to ensure territorial cooperation amongst local actors.

With regard to the apportionment of the financial burdens, in addition to the funding from the 26 municipalities through which the *Olona* river flows, it is important to note that as early as the second program of actions there had also been growing financial support on behalf of private entities such as the *Consortium of the Olona River*, an irrigation consortium that brings together private parties, agricultural and other productive undertakings.

In addition, the case of the *Olona-Bozzente-Lura River Contract* is also noteworthy for analyzing the horizontal and vertical relationships established with existing supra-municipal planning tools that, as already mentioned for the French case study, constitute the plans with which RC are capable of establishing the most significant and worthwhile interconnections.

With reference to urban and territorial planning, the *Olona-Bozzente-Lura* river basin is impacted as well by several instruments drawn up and approved by various local institutions.

The *Regional Territorial Plan (PTR in the Italian acronym)*, approved by Regional Administration in 2010, outlines a strategic vision of development conceived as a frame in which to weave together the planning tools of the municipal level. By virtue of its multi-disciplinary contents, this plan ties in with other planning tools and sectoral policies, in accord with the provisions of Regional Law No. 12/2005 entitled *Legge per il governo del territorio*, i.e. law on territorial government. In particular, by way of the *Regional Landscape Plan* (*PPR in the Italian acronym*), which is a discreet subsection of the *PTR*, the importance of creating synergy between urban and territorial planning and other sectoral planning instruments is also emphasized.

The *PTR* identifies hydrographic basins and sub-basins as the reference units for the orientations regarding the restructuring of the local hydrogeological contexts and requalification of water resources. Out of the Plan emerges a renewed attention for planning at the river basin and sub-basin scales, and the recognition of the need to intervene on river systems in an integrated manner through effective participated policies, regarding strategic planning and negotiated programming applied to each hydrographic units. In this context, the PTR recognizes the key importance of RC, given the current lack of a univocal frame of reference or coordination between urban and territorial, and sectoral planning.

A particular attention was paid also to the relationships between the *Olona-Bozzente-Lura River Contract* and the *Territorial Plans for Provincial Coordination (PTCP in the Italian acronym)*, viewed as the optimal planning level for soil protection as well as for safeguarding water resources, natural environment and landscapes. In line with Regional Law No. 12/2005, *PTCP* are in fact the tools that can ensure vertical integration between the different competences of regional, provincial and municipal administrations and, at the same time, horizontally between the various planning areas.

The *PTCP* of the Province of Milan, approved in 2003, identifies the general direction for planning and development in the provincial territory, with particular reference to issues related to infrastructures, protection of the environment and landscapes, and hydrological and hydrogeological structures. The *PTCP* stresses, in particular, the need to safeguard and enhance the hydrographic network through the integration and coordination of de-pollution and land-take contrast activities, as well as by preserving the quality water resources and ensuring rational groundwater management. Based on the characteristics of the territory, the *PTCP* identified eight *Territorial-landscape Units*, with the further subdivision of the *Watercourse Valley Unit* into the so-called *Olona Valley* and *Valley of Minor Streams* sub-units, comprised of the *Bozzente* and the *Lura*.

The *PTCP of Province of Como*, approved in 2006, set down a number of strategic objectives, which emphasized the need to restore balance between the development of settlements and environmental protection, by leveraging effective coordination among the various territorial policies. The *PTCP* thus offered specific policy orientations for municipal and inter-municipal urban planning, even in terms of improving the system of surface waters, of soil conservation as well as mitigation of hydrogeological risk. Particularly significant is the fact that the *PTCP* reaffirmed the importance of the river basin as a whole, regardless of administrative boundaries, as an incentive to share knowledge and foster concerted efforts among the various players taking part in the exercise of governance over the territory. A ramification of such a mindset is the subdivision of the provincial territory into river basin areas, taken as homogeneous units within which to carry out studies and targeted research projects. The *PTCP* also envisioned support for inter-municipal plans, entrusted with identifying actions aimed at preserving rivers and mitigating river overflows and floods.

The *PTCP of Province of Varese*, approved in 2007, outlined the directions for urban, socio-economic and social development for the provincial territory, contrived into four broad objectives: (I) promoting synergies between education, research and the business sector, (II) enhancing the role of agriculture, (III) fostering development of tourism and territorial marketing, (IV) promoting urban quality standards within its territorial system. The *PTCP* transposes the guidelines contained both in the *Hydrogeological Structure Plan* and the *Water Protection Plan*, and introduces a number of provisions regarding the containment of water consumption and the protection of groundwater resources, which are to be transposed by municipal urban and territorial planning instruments in each local contexts. It should be noted that the *PTCP* endorsed the expectations of the *Olona-Bozzente-Lura River Contract* regarding the limitations and mitigation of flood risk and established that municipalities involved in the application of its plan of actions must take them into account in the drafting of their municipal urban and territorial planning instruments.

With regard to basin planning tools, the *Olona-Bozzente-Lura* territory is impacted by the *Hydrogeological Structure Plan* and the *River Basin Management Plan of the Po River*. In addition, the hydrographic system of the *Olona-Bozzente-Lura* is affected by the *Po River District Management Plan*, drafted by the *Authority of the Po River Basin*, which compounded the provisions comprised within the other tools mentioned above.

Specifically, the *Po River Basin Management Plan* is the programming tool for the safeguard and use of water resources, through which the *Regione Lombardia* has implemented the *Linee di indirizzo strategico per la politica di uso e tutela delle acque*, i.e. strategic policy guidelines for use and protection of waters, developed as of 2002 in line with the European and National regulatory framework, in particular with the EU Water Framework Directive and with Legislative Decree No. 152/1999.

The *Po River Basin Management Plan* is structured in the *Atto di indirizzo per la politica delle acque*, i.e. a programmatic document regarding water policies, and in the *Programma di Tutela e Uso delle Acque*, i.e. the plan for the safeguard and use of waters resources. The latter contains the intervention measures aimed at the integrated protection of the qualitative and quantitative aspects of significant water bodies of the concerned river basin. In particular, on the basis of the matters defined by the said programmatic document, the plan for the safeguard and use of waters resources detailed the state of the art and the objectives to be pursued in terms of surface-water and groundwater quality, in the sense of clarifying the distribution of responsibilities and competences, and thus defining the forms of coordination between different levels of the governance of waters.

The plan for the safeguard and use of waters resources was structured for the individual basins and sub-basins of the concerned regional territory. As regards the sub-basin of the *Olona*, which together with that of the *Lambro* and *Seveso* represent the areas of greatest anthropic pressure, this plan stressed the need to introduce innovative tools, such as RC, for consultation and negotiation processes between institutions and local economic players, so as to share policies and strategies for financial support over the medium and long terms.

4.3 River Contracts in Rural Contexts: The *Basse Vallée de l'Ain* and the *Val d'Ofanto* Case Studies

The two case studies selected for the analysis of river contracts applied to rural settings are the *Contrat de bassin de la Basse Vallée de l'Ain* and the *Val d'Ofanto River Contract*. Specifically, the first case refers to the lower valley of the river basin of the *Ain* river, located in eastern France, in the *Région Rhône-Alpes*. The second case study makes reference to the entire hydrographic basin of the Ofanto river, located in southern Italy, between the *Campania*, *Basilicata* and *Puglia Regions*.

Notwithstanding the diversities in territorial dimensions and the current respective states of these two case studies, the comparison between the *Contrat de bassin de la Basse Vallée de l'Ain* and the *Val d'Ofanto River Contract* is of particular interest to more fully understand the scope of RC applied to those territories with a strong rural connotation and, at the same time, denoted by significant naturalistic value. Consequently, in this comparison, due consideration was necessarily given to their cultural, legal and socio-economic diversities, but also to their respective pathways of activation, which are elements of equally important analytical and informative relevance.

4.3.1 Contrat de bassin de la Basse Vallée de l'Ain

Geographically and from an administrative point of view, la *Basse Vallée de l'Ain* is comprised in the southern part of the *Département de l'Ain*, in the *Région Rhône-Alpes* (Fig. 4.7). The geographical area concerned extends, from north to south, for about fifty km up-stream of the confluence with the *Rhône* river, and is located at 30 km east of the city of Lyon. The *Contrat de basin de la Basse Vallée de l'Ain*, in force from 2006 to 2011, involved a hydrographic sub-unit well defined in terms of its geographical, administrative and socio-economic features, encompassing 40 municipalities with a combined population of about 60,000.

In the mid-part of its course, the Ain river cuts across deep gorges while its natural flow is controlled by five dams (Bravard et al. 1991). This river is one of the best preserved river corridors in the Rhône hydrographic basin, as well as an area of great tourist appeal, due to its environmental riches and landscapes, but also for its accessibility especially from the metropolitan area of Lyon. Along with the *Regional Park of the Haut-Jura* and the reservoirs behind the five dams, the scenery of the *Ain* river constitutes one of the main naturalistic attractions of the entire basin of the Rhône (Dupont 1991).

Of extreme significance is also the immense patrimony in terms of groundwater, which has a strategically prominent role for the entire region. Although there are many water uses quite diversified, these hydrological resources are primarily destined for agricultural irrigation.

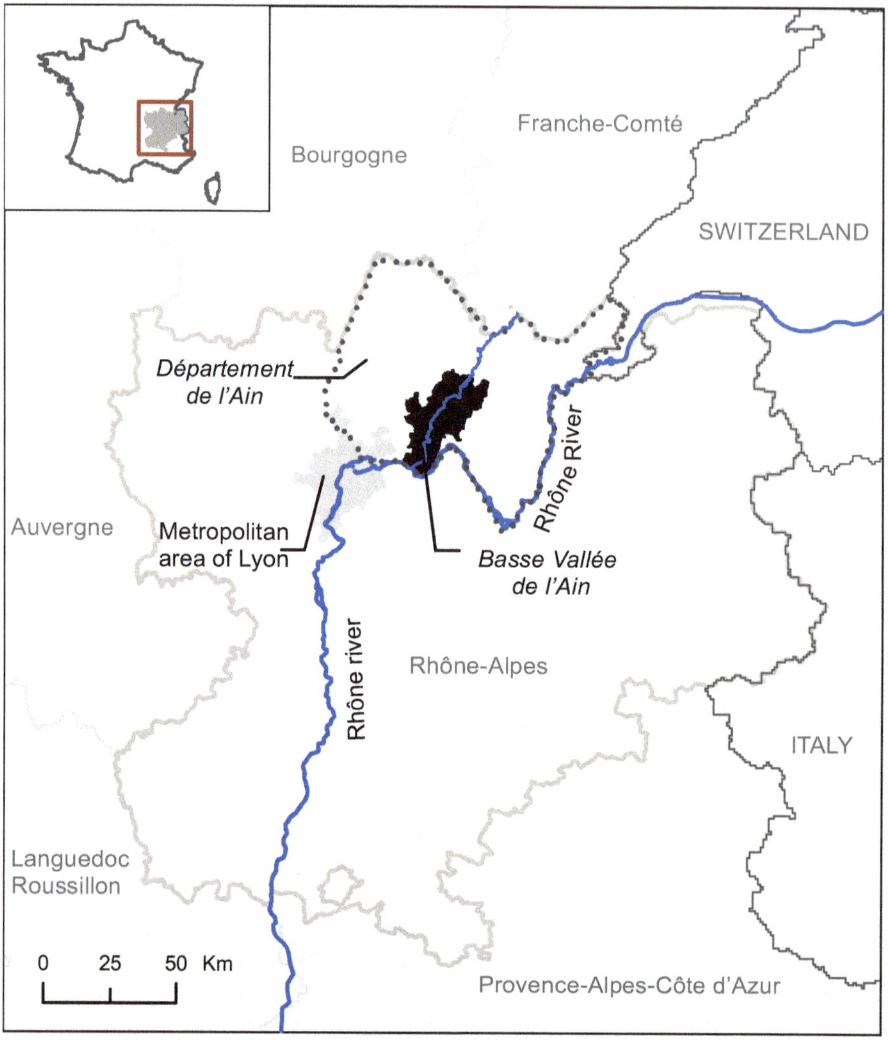

Fig. 4.7 The Basse Vallée de l'Ain in the regional context of Rhône-Alpes

Precisely these agricultural productions, particularly the intensive cultivation of maize, which is the main economic activity of the area, have had a considerable impact in terms of the quality and quantity of water resources and biodiversity (Horizon Centre-Est 2000). In particular, the mosaic of natural environments, as exemplified by the so-called *brotteaux* i.e. the fluvial species extended along the river meanders, constitutes a clear case in point of naturalistic resources still present in this valley, despite the effects caused by intensive agriculture practices of non-traditional crops, by pollution and environmental degradation, also determined by mining activities. The latter, along with the industry of hydroelectric-power production, have had negative repercussions on the overall quality of local landscapes and groundwater, as well as on the natural dynamics of the river.

Notwithstanding the main connotations of the area being decidedly agricultural, it should be mentioned that it also comprises the two highly urbanized areas of *Ambérieu-Pont d'Ain* and *Meximieux-Pérouges-St.Vulbase*, besides the *Parc industriel de la Plaine de l'Ain*. The latter represents the largest and most important industrial center of the Région Rhône-Alpes, among other things, located near to the *Bugey* nuclear power plant (Fig. 4.8).

Specialized reports produced in the early 1980s by the *PIREN-Rhône Group* had already highlighted the naturalistic, environmental and landscape value of the *Basse Vallée de l'Ain*, indirectly helping to put off the planned construction of one of the dams (Bravard et al. 1990; Bravard 2011).

Since the late 1980s, local public actors had been advocating reflection on the topic of urban and territorial planning, as well as on the more specialized issues of water management, followed by a host of initiatives in order to resolve priority environmental challenges and to identify appropriate solutions to address the concerns of economic development within the territory. In particular, the onset of the process dates back to 1987, in the wake of alarming phenomena of massive fish kill that occurred in the summer of the previous year. On that occasion, various public and private actors from the fishery sector and other productive activities joined forces to constitute a first consultation seat with the primary objective of solving the ecological emergency then at hand.

In 1990 the *Conseil Général de l'Ain* eventually instituted the *Comité de Pilotage* for the formulation of an *Etude de définition of a Global Schéma de Gestion de la Basse Vallée de l'Ain*. This initiative had stemmed from certain reflections on the competing uses of water resources and as a result had been oriented to the resolution of the main conflicts, through a program of coordinated

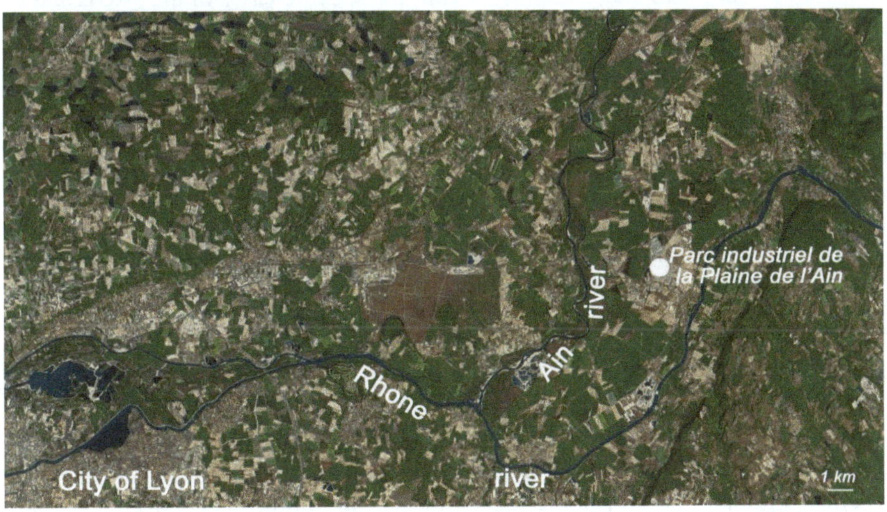

Fig. 4.8 The Ain river at the confluence with the Rhône river, in the north-eastern part of the metropolitan area of Lyon (*basemap* Microsoft Bing)

actions for the benefit of the integrated management of the *Basse Vallée de l'Ain*. Several prior studies conducted in the late 1980s by the *Université Jean Moulin* and by the *Agence de l'Eau Rhône Méditerranée Corse* (Bravard et al. 1991; Dupont 1991), served as stepping stones along this path. Following this first season of specialized studies focusing on the dynamics of the physical and biological characteristics of the valley, its territorial organization and the interdependencies between the diverse uses of water resources, a second phase of local consultation climaxed in the definition of guidelines for a general plan of integrated management. The issues at hand concerned (I) the renaturation of those areas most subject to environmental degradation and the management of natural areas; (II) the analysis and planning of new territorial structures; (III) the protection, development and reorganization of the various uses of water resources. The above investigations, also spurred on by the enactment of the *Loi sur l'Eau* of 1992, were conducive to a feasibility study for a *Schéma d'Aménagement et Gestion des Eaux* (*SAGE*). The perimeter of the instrument was, however, restricted to the *Basse Vallée de l'Ain* for reasons pertaining to management aspects (Dupont 1991), but also due to divergences of views among local actors, and in particular to the stance adopted by *Electricité de France* (*EDF*), the main national public company for the production and distribution of energy.

This first season of investigations and consultation activities amongst local actors was just the starting point of a ten-year path, between 1996 and 2006, marked by a parade of tools for urban and territorial planning and water resources management. In particular, one should mention the *Contrat de Développement Global*, promoted and funded by the *Région Rhône-Alpes* between 1996 and 1999, the *Schéma d'Aménagement et de Gestion des Eaux* and the *Schéma de Cohérence Territorial* (*SCOT*), both drawn up starting around 1998–1999 and ultimately approved in 2002 and 2003, respectively. In the specific cases of the *SAGE* and the *SCOT* there was an actual integration between the two planning tools, already at the programming stages. From the start, the activities of both working groups were, in fact, characterized by the sharing of reflections and knowledge as well as objectives and program orientations, for the cause of integrated management of water resources and of the territory (Semelet 2005).

Following the trail of this wide-ranging and intricate path, it lead up in 2006 to the signing of the five-year action program of the *Contrat de bassin de la Basse Vallée de l'Ain*, whose perimeter matched the one already identified by the SAGE. Its perimeter practically coincided with the administrative limits of the 40 municipalities that had already banded together by way of the *Syndicat intercommunal de la Basse Vallée de l'Ain*.

The latter was responsible for mapping out the knowledge framework regarding the Ain river basin, which was the groundwork for identifying the main aims of the *Contrat de bassin*, largely focused on its specific agricultural needs and those pertaining to hydroelectric-power production. In line with the guidelines of the *SAGE* and with the directions set forth by the European Water Framework Directive, the *Contrat de bassin* identified nine primary objectives and 95 actions, organized into five thematic sections, so called *volets*: (I) improvement and preservation of water-quality status; (II) environmental restoration, management and enhancement of the quality standards of natural environments; (III) prevention and protection against the risk of erosions and flooding; (IV) optimization of the quantitative dimensions of water resources, especially with regard to drinking water; (V) coordination, augmenting public awareness and monitoring the action program. Specifically, given the predominantly agricultural connotations of the area, many actions of the program were aimed at reducing pollution produced by agricultural cultivations, particularly in the areas serving as reservoirs for drinking water, and at raising awareness among farmers of more sustainable agricultural practices. In fact, many consultation activities directly involved representatives of the farming community, also through the promotion of a *Charte des bonnes pratiques agricoles* and a specific program agreement between the *Syndicat de la Basse Vallée de l'Ain* and the farming community at large.

The *Contrat de bassin* was co-signed in 2006 by the *Préfet* of the *Département de l'Ain* on behalf of the State, by the *Agence de l'Eau Rhône-Méditerranée-Corse*, the *Région Rhône-Alpes*, the *Département de l'Ain*, the *Departmental Federation of Fisheries and Aquatic Environment Protection of the Ain*, the *Syndicat intercommunal de la Basse Vallée de l'Ain*, the *Conservatoire Rhône-Alpes des Éspaces Naturels* and also by the public company *Electricite de France*.

Of all the players involved, the key role was that of the *Syndicat intercommunal de la Basse Vallée de l'Ain*, established in 1998, which had already drawn up the *SAGE*. The *Syndicat* actually assumed the function of *structure porteuse* of the *Contrat de bassin* as well as the role of technical and administrative secretariat for the *Commission Locale de l'Eau* (*CLE*). The latter replaced the *comité de rivière*, since in this specific case the perimeter of the *Contrat de bassin* coincided exactly with that of *SAGE*. Furthermore, the *Syndicat* directed the implementation of the plan of actions and ensured coordination and consultation among all partners, besides developing the financial plan and redefining it during the implementation phase.

It is interesting to note that in its capacity as technical secretariat of the *CLE*, the *Syndicat de la Basse Vallée de l'Ain* also assumed the function of ascertaining the compatibility between the procedures of approving or revising the SCOT and the *Plans Locaux d'Urbanisme*, with respect to the guiding principles defined in the SAGE. In this sense, the *Syndicat* proved capable of contributing to a better coordination of urban, territorial and sectoral planning tools. In this regard, it is deemed important to highlight that the *Contrat de bassin de la Basse Vallée de l'Ain* was implemented within a territorial context in which several inter-municipal bodies, other *syndicats* and various associations in the agricultural, fishing and tourism sectors were already vested with specific competences. Some of the latter bodies, in fact, assumed the role of *structures porteuses* regarding plans and programs for local development, promoted in collaboration with the regional administration, being equally oriented towards safeguarding and re-qualifying natural environments and preserving water resources.

Throughout the activation phases of the program of actions, the *Commission Locale de l'Eau*, aside from the fundamental function of constituting the arena for territorial consultation, took on the burden of periodically assessing the progress made, validating the program and proposing changes to the plan of actions, whenever deemed necessary.

As for funding, approximately half of the total costs of the *Contrat de bassin de la Basse Vallée de l'Ain* were borne by the State, through the *Ministry for Ecology, Sustainable Development and Energy*. Financial support was also guaranteed by the *Agence de l'Eau Rhône Méditerranée Corse*, the *Région Rhône-Alpes*, the *Conseil General de l'Ain*, the *Fédération de Pêche de l'Ain* and the company *Électricité de France*.

As done for the other case studies analyzed, due consideration was given to the analysis of the horizontal and vertical relationships that the *Contrat de bassin de la Basse Vallée de l'Ain* established with the other planning instruments in force at the supra-communal level. As already recalled, the decision to make specific reference to the latter level of planning stems from the observation that plans at the supra-communal scale are precisely those that most closely relate to the reference unit of the hydrographic basin, given the territorial contexts for which they were designed.

With regard to the instruments of urban and territorial planning, the *Basse Vallée de l'Ain* is impacted by the *Directive Territoriale d'Aménagement* (DTA) *de l'Aire Metropolitan Lyonnaise*, by the *Schéma de Cohérence Territoriale Bugey-Côtière-Plaine de l'Ain* and, only to a minimal extent, by the *Schéma de Cohérence Territoriale Dombes*.

As mentioned in the paragraphs dedicated to the *Contrat de Rivière de l'Yzeron*, the *DTA de l'Aire Métropolitaine Lyonnaise*, approved in 2007, concerns the vast territory of the three agglomerations of *Lyon*, *Saint-Etienne* and *Nord-Iseère*, some urban centers and a host of smaller rural centers, as well as areas of agricultural and natural significance. In matters of water resources management, the *DTA* makes explicit reference to both surface and underground waters within the framework of its objective to advance policies promoting the conservation and enhancement of the quality status of natural and agricultural areas, aiming at the creation of an integrated system inclusive of these functional domains.

This thematic objective of the *DTA* undoubtedly represents a telling point that is shared with the *Contrat de bassin de la Basse Vallée de l'Ain*, in particular as applies to flood control, improving the quality status of landscapes and, above all, safeguarding standards for the quality and quantity of water resources, especially pertaining to drinking water.

Also noteworthy is the fact that the *DTA* produced ample evidence of striving for integration between sectoral and urban and territorial planning policies, while pursuing quality institutional relationships between the State and local authorities. To this end, the DTA stressed the need for more effective integration between the *Plans Prévention Risques Inondation* and the tools of integrated management of water resources represented by *SDAGE*, *SAGE* and *CdR*.

The same reference area of the *Contrat de bassin de la Basse Vallée de l'Ain* was also affected by the *SCOT Bugey-Côtière-Plaine de l'Ain*, approved in 2002, which covered the administrative territory of 85 municipalities within the river basin.

Within the general objective of balanced development, from the territorial perspective, this SCOT aimed at promoting more sustainable forms, identifying the guiding principles able to restore and maintain an appropriate ecological and socio-economic equilibrium among the various functional domains.

With reference to the territorial scope for which the *SCOT* was devised, the Ain river was viewed as the backbone that actually holds together the urbanized, rural and natural areas. In this sense, the *SCOT* drew on the objectives of the *SAGE*, reaffirming the need to promote farming practices proving more compatible with dynamics of the river, with the aims of environmental requalification and the need to safeguard water resources, thus taking a number of its cues from the *Contrat de bassin de la Basse Vallée de l'Ain.*

With regard to river basin planning tools, the *Basse Vallée de l'Ain* is impacted by the *SDAGE Bassin Rhône-Méditerranée* 2016–2021 and the *SAGE de la Basse Vallée de l'Ain.*

As reported for the case study of the *Contrat de Rivière de l'Yzeron*, the SDAGE 2016–2021 reaffirms the main guidelines of protecting and enhancing the quality status of water environments at the basin scale, as defined by the previous *SDAGE 2010–2015*. The declared priority of the latter plan was that of achieving a *good status* of water quality by 2015, in compliance with the dictates of the European Water Framework Directive.

In particular, the *SDAGE* tend to bolster local management of water resources and to ensure coherence between urban planning and water management within the territory, underscoring the need to ensure and strengthen the coherence among planning tools and build dependable relationships between those institutional actors directly responsible and every local stakeholder.

In close analogy to the case illustrated for the *Yzeron* basin, even the *Contrat de bassin de la Basse Vallée de l'Ain* found various policy congruities with the *SDAGE 2010–2015*, especially in the thematic section devoted to finding measures for reducing sources of pollution of agricultural origin, curbing patterns of water overconsumption and managing hydraulic works.

Getting down to the specifics of the issues concerning pollution caused by agricultural practices, it is interesting to note that within the *SDAGE*, the *Basse Vallée de l'Ain* was listed among those areas targeted for specific programs of actions, pursuant to the EC *Nitrates Directive*. Indeed, it was classified among those sub-basins requiring measures to curb emission levels of pollutants resulting from the use of certain pesticides, so as to improve the quality standards of their waters.

The *SAGE de la Basse Vallée de l'Ain*, approved in 2003, constitutes a planning tool developed on local actors own initiative so as to identify general objectives for water resources management that take into account the different uses of water resources.

The main issues identified and addressed within the *SAGE* framework made reference to: (I) the anthropogenic altering of the natural dynamics and hydrological regime of the Ain river, especially caused by hydroelectric-power production; (II) erosion phenomena and risk of floods; (III) eutrophication phenomena and protection of fish fauna; (IV) the generalized degradation of natural environments; (V) the state of strife from conflicting uses of water resources; (VI) the underexploited potential of the territory as regards tourism.

In this perspective, the SAGE clearly highlighted the importance of integrated management of surface and underground waters, as well as of the processes of territorial consultation and participation, revealing in that sense a significant degree of osmosis, on several points, between the same *SAGE* and the program of actions of the *Contrat de bassin de la Basse Vallée de l'Ain*.

4.3.2 Val d'Ofanto *River Contract*

The case study of the *Val d'Ofanto River Contract*, albeit still in the making, is an example of utmost significance, inasmuch as it concerns a vast hydrographic basin overlapping part of the three regions of *Campania*, *Basilicata* and *Puglia*, in southern Italy (Fig. 4.9). In fact the *Ofanto* river marks the administrative limits between the Provinces of *Avellino* (*Campania*), *Foggia* and *Barletta-Andria-Trani* (*Puglia*), as well as *Potenza* (*Basilicata*), covering an area comprising 51 municipalities, with a combined population of about 420,000 inhabitants.

The *Ofanto* river is the most important watercourse in the Region of *Puglia* and it flows for approximately 170 km, with a total hydrographic basin of about 2670 km^2, among the vastest of southern Italy. Along its course the Ofanto is fed by many tributaries consisting mostly of small seasonal streams and some torrents. In its up-stream sections the riverbed is narrow and steep, whereas proceeding down-stream it flows through wide valleys with a flat-bottomed riverbed (Russo 1998).

Especially in the context of the *Regione Puglia*, the *Ofanto* river has deeply influenced the settlement patterns of the area, also historically representing a foremost factor in the ecological as well as anthropic interconnections between areas situated inland and along the Adriatic coast (Fig. 4.10).

The territory of this basin is mostly characterized by rural landscapes and some areas of outstanding naturalistic significance. Moreover, several industrial centers are also present in the upper *Ofanto* valley. The considerable abundance of groundwater, particularly in the stretch spanning the *Regione Puglia*, whose geomorphology is characterized by permeable rocks and karst phenomena, is in stark contrast to the marked scarcity of surface water resources (Barbanente and Monno 2005). This problematic situation has been accentuated by the process of industrialization, linked in large measure to the reconstruction that ensued in the wake of the earthquakes of 1980, as well as to the growing demand due to intensive agricultural practices and, in general, to an approach to water resources management geared almost exclusively to water catchment for the benefit of other productive activities (Scognamiglio 2004; Barbanente and Monno 2005). Adding to the above, other critical factors have appeared over time, namely high anthropic pressures and the haphazard dynamics of an urban expansion almost always devoid of any unitary strategic vision.

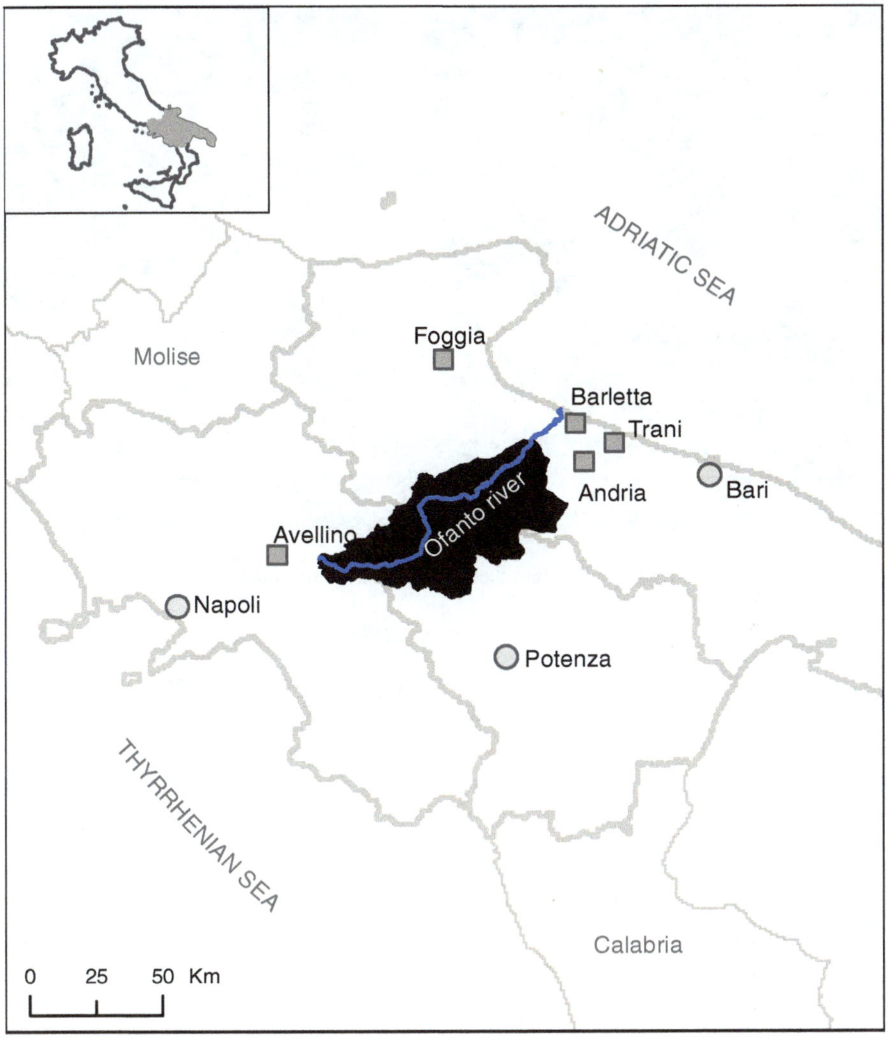

Fig. 4.9 The Ofanto river basin in the interregional context of Campania, Basilicata and Puglia

By and large, the lack of coordination that has characterized the forms and modalities of water resources management in the *Ofanto* basin have severely compromised actions to safeguard and preserve fluvial contexts, also due to the considerable fragmentation of management competences among a host of local authorities and other bodies present in the territory.

In sum, the main water-related issues taxing the *Ofanto* Valley are: (I) the radical and irreversible modifications to the hydrological and geomorphological characteristics of the river; (II) the disruption of the ecological equilibrium and natural water regimes due of the intense water catchment practices; (III) the anthropogenic

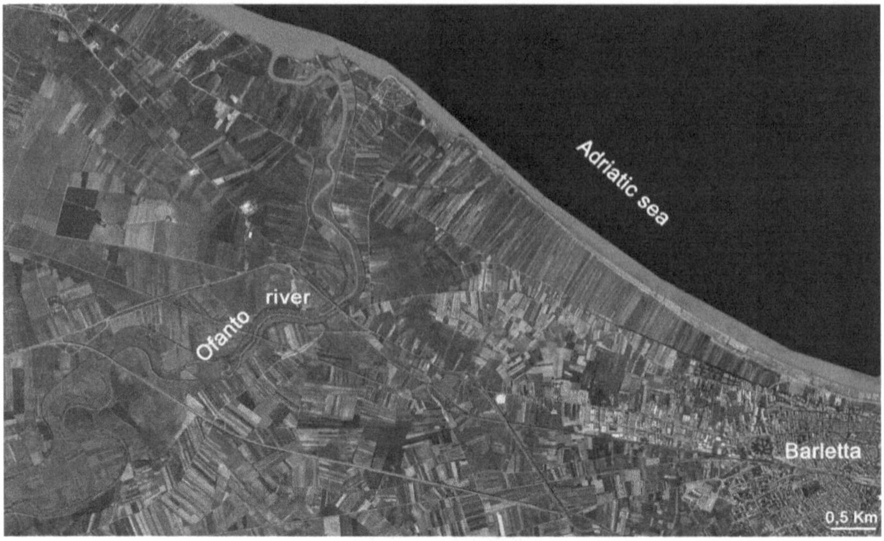

Fig. 4.10 The Ofanto river flowing into the Adriatic sea, north to the urbanized area of Barletta *(basemap* Microsoft Bing)

pressures deriving from its settlement patterns as well as from agricultural, industrial and mining activities that have determined high levels of pollution within the fluvial context; (IV) the increased risk of desertification of the coastal plain.

In order to address these issues, with particular regard for the absence of strategies shared by all local stakeholders, a number of programs and initiatives aimed at fostering development within the territory have been launched since 2002.

Such projects, by promoting new forms of cooperation between the public and private sectors, as well as by sharing knowledge bases and strategic policies, have opened the way to new and more inclusive forms of territorial consultation, relying on renewed inter-disciplinary and participated approaches (Barbanente and Monno 2005).

Based on these premises, in the second half of the 2000s a process of cooperation was launched, involving the *Regione Puglia*, municipalities and farmers, so as to promote the creation of the protected natural area of the *Ofanto* River, instituted by a Regional Law in 2007.

Subsequently, institutional procedures, development and promotion programs targeting local areas as well as new specialized studies all followed. With these as stepping stones, this pathway ultimately peaked in the debut, in 2009, of the so-called *River Pact for the Ofanto Valley*. The latter is an interregional framework program promoting novel prospects for the integrated development of the territory of the *Ofanto* valley, drawing on the concept of *bioregionalism* (Saragosa 2005; Magnaghi 2011), thus pursuing greater integration between natural, anthropic and economic systems (Iacoviello 2011). This first agreement among the three regional administrations of *Basilicata*, *Campania* and *Puglia*, the four Provinces of *Avellino*,

Barletta-Andria-Trani, *Foggia*, *Potenza*, and 51 interested municipalities, represented the start of the actual consultation phase. The purpose of the agreement was to promulgate the *Declaration for the River Pact of the Ofanto Valley*, which aimed to enhance the distinguishing features of the territory and increase the awareness on the part of the local communities, regards the need to embark upon and implement shared and participated initiatives to promote sustainable development, at the scale of the entire hydrographic basin.

It is important to highlight the acknowledgement of the role of the *ecological network*, also in the initial phase, in terms of river basin dynamics, by way of their systematized development, inasmuch as they were already included amongst the planning instruments approved by the provincial administrations.

Even aspects regarding the restoration of the historical fabric of inhabited areas along river contexts found their place in the drafting of the declaration, with the overarching objective of developing a well-balanced relationship between urban and environmental systems (Iacoviello and Scaduto 2012). Specifically, these initiatives found direct correlation with the substance of the *Regional Landscape and Territorial Plan* (*PPTR in the Italian acronym*) of the *Regione Puglia*, in which the *Ofanto* River was designated an urban and territorial park.

In 2014 interregional cooperation was resumed with the signing of a new agreement dubbed *Pact of the Ofanto valley—Declaration for Sustainable Interregional Development of the Ofanto valley in European Planning 2014–2020*, always from the perspective of more unified territorial development in the *Ofanto* valley, beyond the mere limits of individual institutional competencies. This statement of intent was promoted and cosigned by the Province of *Barletta-Andria-Trani*, several municipalities of the Provinces of Avellino, Foggia, *Barletta-Andria-Trani* and *Potenza*, the *Regional Natural Park of the Ofanto river*, the *Territorial Environmental Agency* of the Province of *Barletta-Andria-Trani*; the *University of Bari*, and also some irrigation consortia.

The Declaration of 2014, following in the footsteps the previous declaration of 2009, defined among its priority objectives that of improving the coordination between the public and private sectors at the interregional scale of the Ofanto river basin, in order to pursue new lines of sustainable territorial development, also through National and European funding, according to the 2014–2020 EC programs. The document also reaffirmed the tenet of pursuing new forms of integrated water resources and fluvial-environmental management, also promoting and broadening affiliation in the *Val d'Ofanto River Contract*. To this effect, it remarked on the need for a negotiated agreement for the entire interregional hydrographic basin through an integrated and interdisciplinary approach, so as to overcome the fragmentation of actions that had thus far characterized the efforts of local actors.

The Declaration of 2014 re-launched the importance of interregional governance over the *Ofanto* valley as well as of implementing actions in line with the objectives and regulations of national and European programs for the period 2014–2020.

This statement of intent was initialed in the form of a document open to subsequent subscription on the part of other public and private entities, concerned with the dynamics of territorial development.

With regard to the institutional players unaccounted for among the signers of the Declaration of 2014, the absence of the different Basin Authority responsible for the Ofanto Valley is particularly noteworthy. This facet evidently reveals how, notwithstanding the worthy efforts made in preparation of the signing of the contract unwavering perseverance is undeniably going to be required on the part of the promoters for the actual stipulation of the *Val d'Ofanto River Contrat*. In practical terms, the priority arguably remains the level of fragmentation of the competencies and roles of the various institutional players, despite attempts made to clear the hurdle via the various programs of actions, initiatives and, even, through the very Declarations of 2009 and 2014. This appears to be the decisive prerequisite, granted all actors involved actively cooperate, in order to put into effect a truly integrated water resources management at the river basin scale.

The *Val d'Ofanto River Contrat*, albeit still awaiting to be formally activated, points to potential interactions with the currently available tools for urban and territorial planning with regard to the basin territory.

Among the landscape and territorial plans, and the spatial planning schemes at the supra-municipal level, those certainly worth mentioning are the *Regional Landscape and Territorial Plan of the Regione Puglia* (*PTPR in the Italian acronym)* and the *Provincial Territorial Coordination Plans (PTCP in the Italian acronym) of Avellino, Foggia, Barletta-Andria-Trani* and *Potenza*.

The former, approved in 2015, aims to enhance landscape identity while achieving more appropriate levels of sustainable development within the concerned territory. The plan explicitly highlights the role of negotiated programming tools and participated forms of governance and, in particular, those of *territorial pacts* and local RC. Specifically, the Implementation Technical Standards of the PPTR state that «[…] the Region promotes RC, […] negotiated programming tools directed to the adoption of a system of shared objectives and rules, by means of consultation processes and the integration of actions and projects permeated with the culture of water as part of the common good […]». All the more, both local territorial pacts and RC have been comprised within the PPTR under the heading of experimental integrated landscape projects.

With regard to the 12 general objectives of the PPTR, the major points of congruence with the concept of the *Val d'Ofanto River Contract* emerge with reference to *Objective 1—Achieve a state of hydrogeomorphological equilibrium within the river basin*. In fact, this objective emphasizes how the pursuit and maintenance of a stable and resilient hydrogeomorphological equilibrium within the *Ofanto* hydrographic basin and sub-basins is deemed the crucial precondition of any effective spatial and landscape planning.

It should be stressed that in relation to the specific hydrogeological structure and geomorphology of the *Val d'Ofanto* territory and, therefore, of the foreseeable scenario of the RC, the *PPTR* outlines two major intervention strategies: (I) in the upper valley, the ecological-naturalistic requalification of rivers and the definition of a system of ecological multi-functional corridors; (II) in the mid and lower valley, the hydraulic requalification of karst groundwater drainage systems, with particular regard to issues concerning urban waste-water treatment. Moreover, given the fact that some of the specific *PPTR* objectives entail incentives for more environmentally-friendly agricultural practices, especially in terms of curbing water overconsumption, presumably further margins of interplay with the *Val d'Ofanto River Contract* will likely result.

The *PTCP*, drafted and approved in recent years for the provincial areas comprised within the *Ofanto* hydrographic basin, represent the general programming and strategic policy tools at the supra-municipal level. In this perspective, they strive to ensure the coordinated development of provincial territories while pursuing the general aims of: (I) safeguarding and promoting of rural areas and landscapes; (II) preserving natural resources and maintaining historical settlement patterns; (III) defending soil-quality status and hydrogeological structure; (IV) promoting local economic networks, and bolstering and interconnecting infrastructure of supra-municipal significance, service networks and mobility systems; (V) coordinating and orienting municipal urban planning instruments.

In promoting the development of ecological networks, the PTCP support the systematizing of ecological networks and corridors, within which a recognized and important role is played by rivers and their natural dynamics. In light of these overarching guidelines, the *Val d'Ofanto River Contract* will likely find many points of congruence as well as many opportunities for integration with the same PTCP.

With regard to basin planning, the existing instruments for the concerned territory are represented by the *Piano per l'Assetto Idrogeologico* (*PAI in the Italian acronym*) for each region of *Campania*, *Basilicata* and *Puglia*, and the *Regional Water Protection Plans* (*PRTA in the Italian acronym*).

The analyses, information and specifications of these sectoral plans merged into the *Basin District Water Resources Management Plan of the Southern Apennines*. This plan was drafted by the *National Basin Authority of the Liri-Garigliano and Volturno Rivers,* and by the regional administrations, pursuant to the European Water Framework Directive and its Italian transposition, represented by the Legislative Decree No. 152/2006. With specific reference to the territories of the river basin of the *Ofanto,* the *Basin District Management Plan* highlights the numerous critical environmental issues, particularly with regard to the impact of intensive farming practices and the use of chemical fertilizers and fungicides. It is worth noting that within this Management Plan there is a specific reference to RC, acknowledged as means of concerted programming and planning for the integrated protection and management of water resources, soil and the environment. This attribute leads to further prospects for integration with the awaited *Val d'Ofanto River Contract* and, in general, with the superordinate tools for basin planning.

4.4 Synthesis of the Comparative Analysis

The four case studies analyzed allow one to draw initial comparisons between the application of river contracts in France and Italy. Similarities and differences between the French and Italian scenarios are apparent, both from a broader perspective and on detailed observation of the forms taken on by these agreements with reference to hydrographic basins concerning metropolitan areas, on the one hand, and rural areas, on the other.

Specifically, the comparative investigation conducted on the two national scenarios through the four case studies considered, comprehensive of direct interviews with the respective promoters, warrants several general observations on the regulatory, procedural, and operational aspects as well as evaluations regarding contents and participation.

In terms of regulation, the French legal and procedural background of *contrats de rivière* is more articulate and well-endowed with specific regulatory references. In fact, it emerges how policy-wise, the French track record boasts over thirty-years evolution in its experience as regards integrated and participated water resources management. In this context, the regulatory framework for *CdR* were defined by way of a number of circulars issued by the Ministry of the Environment between 1981 and 2004, which outline the specifics of its regulatory structure in terms of procedures and contents.

In Italy, the more limited experience with river contracts, introduced only recently in comparison with the French reference model, has yet to develop an explicit legislative and regulatory paradigm valid at the national level. In fact, besides propositional documents, such as the *National Charter of River Contracts*, or a number of bills under consideration, promoters of RC have no specific, clearly defined regulatory references anywhere resembling the French ministerial circulars. Therefore, the actual procedures to be followed in initiating and implementing RC in Italy are defined on a case-by-case basis according to regulation and policy documents or guidelines drawn up, only of late, by individual regional administrations.

A legislative analogy between the two contexts of France and Italy, however, appears in national regulation providing orientations for local policies concerning integrated water resources management. In fact, the respective laws transposing the European Water Framework Directive (WFD) of 2000 acknowledge the river basin as the unit of reference for the optimal implementation of these policies. This regulatory common denominator constitutes a pivotal element for the implementation of RC as well as for their integration with the instruments of urban and territorial planning.

Another similarity between France and Italy is the impact of the WFD on the diffusion of the RC paradigm in each national context. In fact, the transposition of this EU directive gave renewed impetus to France, which occasioned a second generation of *contrats de rivière*, and spurred the first season of experimental experiences in Italy.

As concerns the integration between the tools of water resources management and those of spatial planning, another analogy between the two regulatory frameworks can be elucidated. In France, the same Law No. 338/2004 that transposed the WFD, also sanctioned the prerequisite of full compatibility among urban planning instruments issued at every level, with reference to the guiding principles of integrated water resources management as defined by the *Schémas Directeur d' Aménagement et de Gestion des Eaux* (*SDAGE*) and by the *Schémas d'Aménagement et de Gestion des Eaux* (*SAGE*). Similarly, Legislative Decree 152 of 2006, which transposed the WFD into Italian legal framework, introduced the requirement of adapting local urban planning instruments according to the orientations of Basin Plans, entrusting regional administrations with the burden of monitoring the matter. In this regard, it should be noted that some of these administrations, through their spatial planning laws and instruments, have helped to promote the integration of urban and sectoral planning, in some cases availing themselves of RC.

With regard to the procedures and matters pertaining to the launch and implementation of RC, the comparative analysis highlights further similarities and differences.

With reference to the length of the procedures required in order to initiate a RC, it is common knowledge that equally in France as in Italy, the time-frame needed is on average approximately ten years. As seen throughout the analysis of the four case studies, the reasons behind this time factor are related, on the one hand, to the ecological and environmental emergencies that arise and the time required for the preparation of the ensuing knowledge-mapping and specialized studies. On the other hand, the length of the preparatory phase is conditioned by the time-consuming and laborious process of setting in motion the dialogue and consultation among a host of parties involved, as well as the need to find solutions that are adequately modeled on the scale of the basin territory.

In all RC experiences, the undeniable importance attached to the mapping out of the general knowledge base, prerequisite to defining the actual program of actions, represents an element whose relevance is two-fold, firstly with regard to the scientific and technical contents, and secondly in terms of stakeholder participation. In fact, the piecing together of a common knowledge base offers institutions and other public and private organizations the opportunity to exploit prior studies and research, while integrating them into a common framework that takes into due consideration all the needs and viewpoints of the various participants.

As for the approval procedure and actual launch of RC, the French scenario presents a more detailed definition of the elaboration phases and characterization of the role of the institutional actors responsible for the validation of the action plan. In fact, the preparation stage of the *contrat de rivière* entails two drafts of the agreement and the relative action program, a preliminary and a final version, each drawn up by the promoter and then submitted to the *Comité de bassin*, whose approval is required for validation and subsequent signature of the agreement. In contrast, in Italy the drafting and approval procedures of RC vary from case to case, but usually entirely devolve on the regional administration that, in most cases, is likewise the proponent.

Regarding the programs of actions, the results of the comparative analysis show a marked distinction between the structuring of the French *CdR*, based on predefined *volets* expounded in ministerial circulars, and the relative heterogeneity that characterizes the structuring of RC in the Italian context.

Comparing the proportions of the surface areas to which RC apply, in France these tools generally impact whole sub-basins. Nevertheless, cases such as the *Basse Vallée de l'Ain* attest to the fact that the extension of some *CdR* can on occasion be subject to specific interests or situations of local conflict, with limiting effects. In Italy, in contrast, there is a mosaic of dimensions, ranging from examples concerning vast basins spanning territories overlapping several regions, such as the *Val d'Ofanto River Contract*, to initiatives impacting hydrographic sub-units, or that are even the mere expression of an amalgam of political interests among local administrative bodies, often devoid of any relation, in the hydrographical sense, to a specific basin context.

The comparative analysis between RC applied to hydrographic units that comprise metropolitan areas and those activated in contexts whose connotations are mainly rural, stimulates reflection on the respective priorities of the issues addressed. In the former context, the programs of actions of RC focus more on matters of a hydrogeological nature, on mitigating flood risk and management of drinking water. Vice versa, RC applied in the latter context favor curtailing pollution related to agricultural practices and optimization of water supply systems to the benefit of farm crops. In the specific case of the *Contrat de bassin de la Basse Vallée de l'Ain*, to the above issues one must also add those concerning the interests of the fishing sector, on the one hand, and of hydroelectric power production, on the other. Beyond the specific priorities, a common element to both types of RC is represented by the inclusion of natural reserves and protected areas of natural interest, which, in many cases, effectively serve as vectors of integration between urban and rural systems.

The time requirements of RC action plans implementation represent an element of further differentiation between the French and Italian contexts. In fact, in France the range of allotted time for the *CdR* is preset at the signing of the agreement and averages between 5 and 7 years. In Italy, however, the timetables for the implementation of programs of interventions are quite variable and, as in the case of the *Olona-Bozzente-Lura River Contract*, several programs of actions may succeed each other within the same agreement, over a protracted period.

More generally, it seems that time limits set a priori are oftentimes actually incompatible with the evolutionary dynamics of fluvial contexts and, thus, with any claim to the certainty of achieving both optimal and timely results.

A few considerations also seem in order regarding the relationships among RC, tools for spatial planning and those for river basin integrated management. In France, the relationships between *CdR* and territorial and urban planning instruments stand the test of management plans at the basin and sub-basin scales, such as the *SDAGE* and *SAGE* schemes. In particular, these relationships are based on the recognition of *CdR* as operational tools of the SAGEs, as asserted by the second *Loi sur l'eau* of 1992. In addition, the ministerial circular of 2004 on *CdR* reaffirmed the need for effective consistency of these agreements with *SDAGE* and *SAGE*, all the more given the superordinate nature of the latter with respect to urban and territorial local plans. The *Loi Eaux et Milieux Aquatiques* of 2006 further contributed to such integration by sanctioning the importance of cooperation between relevant actors in the fields of water resources management and urban and territorial planning. In this light, the case of the *Contrat de basin de la Basse Vallée de l'Ain* is regarded as a best practices, due to the advanced and effective level of integration between water resources management and other sectoral planning instruments. At the same time, the *Contrat de Rivière de l'Yzeron* offers corroborative evidence that, absent a *SAGE* in force, the integration of *CdR* and spatial planning tools may prove weak and, thus, less effective.

In the Italian context, the connections linking river contracts, river basin management plans and the tools of urban-territorial planning, still have a degree of definition insufficient to avoid overlap in institutional competences and actions and, therefore, to ensure effective forms of integrated management of land and water resources. In this perspective, the *Val d'Ofanto River Contract*, even if still in making, may be considered a suitable model of a strategic scenario that may serve to orient efforts and actions of public and private players, so as to properly integrate the various tools of sectoral planning.

Both in France and in Italy, promoters of RC are in most cases local authorities and public bodies. Specifically in France, actors that generally take on the role of *structure porteuse* (project manager) of the *CdR* are the *syndicats intercommunaux*, such as in the case of the *Yzeron* and the *Basse Vallée de l'Ain*, which band together as associations of municipalities. In this capacity, the *syndicats* oversee to the preparation of the *CdR* and action programs; moreover, they supervise activities and monitor interventions, besides constantly coordinating and supporting the dialogue amongst partners. In this regard, the promoters see to identifying and organizing appropriate forums for dialogue and participation, as well as setting in motion the means for communication and public information. Despite all effort put into the endeavor, in some cases there is a want of synergy between the actors playing a strategic role, as was the case in the *Contrat de Rivière de l'Yzeron* where the institutional interactions between the *syndicat intercommunal* and the metropolitan city of Lyon were, to some extent, ineffective. Among the other key players whose participation tends to ensure the success of a *CdR*, they may also include the State, the Regional administrations and their agencies responsible for management and protection of water resources, associations of farmers, fishermen and of the industrial sector, environmentalist groups and other local stakeholders. As far as funding is concerned, the State, the *Agences de l'Eau* and the *Régions* are the main financial backers of *CdR*, while contributions on the part of the private sector have dwindled.

In Italy, the role of the promoter is generally assumed by regional and provincial administrations, especially due to the respective competences of these local authorities in the management of water resources and the territory. On occasion, municipalities may also assume the project management role, also as funding partners, as well as *Basin Authorities*, *Ambiti Territoriali Ottimali* (*ATO*), universities, park authorities, trade associations and environmentalist groups.

In some cases, regional and provincial administrations implement contracts impacting river basin that span multiple provincial and regional territories, assuming, therefore, those duties and powers that would seem more befitting of the

Basin Authorities. Indeed, the latter seem those most capable of ensuring ample coordination and cooperation at the interregional level, and could more readily promote an effective integration of RC with the *Basin Plans*, *Hydrogeological Structure Plans* and *Water Protection Plan*.

By and large, in both national contexts analyzed, the stability and durability of the organizational and implementation structure behind the RC appear to be of vital significance. These two factors actually appear to outweigh the type of institution actually entrusted with setting up or representing the operational structure. The features of stability and durability, however, while of considerable importance by itself, are so, especially, to the extent that they are the expression of a strong political facet that in all cases characterizes the promoters of RC. Nevertheless, this political connotation at the same time represents an ever present threat to the stability of institutional cooperation, the continuity of dialogue and, therefore, the achievement of the objectives planned. In this sense, a particularly telling experience is that of the *Contrat de Rivière de l'Yzeron*, in which the administrative elections of 2014 marked a gap in the continuity of local policy on water resources management, such that the start of the *SAGE* and a second *CdR* had to both be deferred. Similarly, the recurrently alternating local political representatives throughout the implementation of the *Olona-Bozzente-Lura River Contract* was undoubtedly a hindrance to continuity regarding institutional vision and, thus, perspective on the overall project design.

The synthesis heretofore presented, contrasting the contexts of France and Italy, provides the basis for further considerations regarding the current framework, in terms of dissemination and application of river contacts, and its foreseeable future scenarios and evolution. These arguments are more fully described in the Chap. 5.

References

Barbanente A, Monno V (2005) Changing discourses, practices and spaces of coexistence: perspectives for the environmental regeneration of the Ofanto River Basin. Plann Theor Pract 6(2):171–190

Bravard J-P (2011) Genealogy of the river "mobility space" concept, France. Theory end implementation aspects. Atti del I Congreso ibérico de resturación fluvial Restauraríos, León, 18–20 Ottobre 2011, pp 46–58

Bravard J-P, Franc O, Landon N. Large JL, Peiry JL (1990) La basse vallée de l'Ain: étude géomorphologique. PIREN-CNRS, Agence de l'eau RMC

Bravard J-P, Malavoi JR, Amoros C (1991) L'Ain, ou la difficulté de gérer une rivière en cours de métamorphose. Actes de la Journée d'Étude du 17 Mars 1989 Rivières en crise. Saône, Ain, Durance, Univ. Jean Moulin, Lyon

Breil P (2007) A focus on periurban development and aquatic system management. Comunicazione al Congresso organizzato dall'Università di Monash, Melbourne, AUS, 02 febbraio 2007

Breil P, Schmitt L, Lafont M, Frisch H (2006) La ressource en eau dans les hydrosystèmes périurbains. Enjeux, gestion, restauration. L'exemple de l'Yzeron. Communication au Séminaire Gé_Eau_Graphies, Université Lyon 2, Bron, 03 mars 2006

BURGEAP (2011) Étude bilan, évaluation et prospective du CdR Yzeron. Lyon

Calori A (2004) Il Contratto di fiume Olona-Bozzente-Lura. L'approccio integrato e negoziale per la pianificazione a scala di bacino fluviale coinvolge enti locali e cittadini nei processi di riqualificazione del territorio. Il progetto sostenibile, n.4

Clerici M, Borasio M, Canti F (2011) I contratti di fiume in Lombardia. In: Bastiani M (ed) Contratti di Fiume. Pianificazione strategica e partecipata dei bacini idrografici. Approcci, esperienze, casi studio. Dario Flaccovio Editore, Palermo

Dupont P (1991) Basse Vallée d'Ain. La nécessite d'une gestion intégrée. Actes de la Journée d'Étude du 17 Mars 1989 Rivières en crise. Saône, Ain, Durance, Univ. Jean Moulin, Lyon

Ferraresi G, Magnaghi A (1992) La Valle del Lambro, Seveso, Olona: da aree ad alto rischio a progetto integrato di valorizzazione territoriale. In: Magnaghi A, Paloscia R (eds) Per una trasformazione ecologica degli insediamenti. Franco Angeli Ed., Milano, pp 57–78

GRAIE-ZABR (2008) Le bassin versant de l'Yzeron: Point sur les travaux de recherche réalisés, en cours et perspectives. GRAIE, Villeurbanne, Lyon

Horizon Centre-Est (2000) Etude hydrogeologique sur le perimètre du SAGE de la Basse Vallée de l'Ain. Chazey/Ain

Hubert G (1988) Approche méthodologique pour l'étude des cours d'eau urbains et péri-urbains: Application au cas de l'Yzeron. GRAIE, Villeurbanne, Lyon

Iacoviello M (2011) dal Patto Val d'Ofanto ad Apulia Fluminun. Programmazione e pianificazione, alla ricerca di una convergenza possibile tra bioregionalismo, interscalarità, irrequietezza e complessità. In: Bastiani M (ed) Contratti di Fiume. Pianificazione strategica e partecipata dei bacini idrografici. Approcci, esperienze, casi studio. Dario Flaccovio Editore, Palermo, pp 461–487

Iacoviello M, Scaduto ML (2012) Il "contratto di fiume" per la pianificazione del sistema insediativo della Val d'Ofanto: tra approcci bioregionali, animazione territoriale e azioni concorrenti, Planum. J Urbanism 25(2)

Lafont M (2006) Gestion de la ressource en eau dans les hydrosystèmes périurbains (GEREHPUR). Cemagref, Lyon Cedex

Lalo A (1986) Bassin versant de l'Yzeron: Caractéristiques, systèmes d'assainissement, type d'urbanisme. GRAIE, Villeurbanne, Lyon

Magnaghi A (1995) Bonifica, riconversione e valorizzazione ambientale del fiume Lambro, Seveso, Olona. Il Manifesto progettuale, Urbanistica Quaderni, Anno I, INU

Magnaghi A (2004) Il contratto di fiume dell'Olona. Contributo presentato alla Semaine franco-italienne/settimana italo-francese (19–23 aprile 2004) "Amenagement du territoire: confrontation d'experiences/Pianificazione e gestione del territorio: esperienze a confronto", organizzato dal Dipartimento di Pianificazione dello IUAV in collaborazione con il Pole de Competence en Urbanisme a Lyon, pp. 122–140

Magnaghi A (2011) Contratti di fiume e pianificazione: uno strumento innovativo per il governo del territorio. In: Bastiani M (ed) Contratti di Fiume. Pianificazione strategica e partecipata dei bacini idrografici. Approcci, esperienze, casi studio. Dario Flaccovio Editore, Palermo, pp 31–44

Meuret B (1988) Rapport sociologique sur la vallée de l'Yzeron. GRAIE, Villeurbanne, Lyon

Piegay H, Cossin M, Saulnier D. SEAG de l'Yzeron du Ratier et du Charbonnières, SEAGYRC, CNRS (2000) Projet de contrat de rivière Yzeron: Étude paysagère du patrimoine naturel et bâti lie à l'eau sur le bassin versant de l'Yzeron

Radojevic B, Breil P, Chocat B (2002) Quantification du rôle de l'urbanisation en particulier et de l'usage des sols en général sur les crues de l'Yzeron depuis quarante ans. Programme RIO Rapport final

Regione Lombardia (2004) Accordo Quadro di Sviluppo Territoriale (AQST)—Contratto di Fiume Olona-Bozzentte-Lura

Regione Lombardia-ARPA Lombardia (2004) Attività di supporto ai processi negoziali "Verso i Contratti di fiume bacino Lambro—Olona". Rapporto primo anno di lavoro: Contratto di fiume Olona-Bozzente-Lura, aprile 2004

Regione Lombardia-IREALP (2010) Atlante delle politiche. Scenario descrittivo-interpretativo per la riqualificazione paesaggistico ambientale e il contenimento del degrado. Capitolo II—Ambito vallivo Olona, Giugno 2010

Russo R (1998) Ofanto, fiume di Puglia. L'idrografia, la storia, l'ambiente. Editrice Rotas, Barletta

SAGYRC (2007) Aménagements hydrauliques de protection contre les inondations du bassin versant de l'Yzeron. Etude des solutions alternatives, mai 2007

SAGYRC (2008) Présentation du Syndicat de rivière et de la démarche de gestion des cours d'eau. Conseil syndical, 16 avril 2008

Saragosa C (2005) L'insediamento umano. Ecologia e sostenibilità, Donzelli, Roma

Scognamiglio R (2004) Transboundary management of water resources: the italian experience in Basilicata and Apulia. Contributo presentato all'International FORUM on Food Security under Water Scarcity in the Middle East: Problems and Solutions, Como, 24–27 novembre 2004

Semelet J (2005) Le Schéma d'Aménagement et de Gestion des Eaux de la Basse Vallée de l'Ain. Actes de la Conférence Euro-africaine «Eau et Territoires» 22–23 Mars, Unesco Paris

Thollet F, Branger F (2009) Recueil de données du bassin versant de l'Yzeron. Année 2008, Cemagref, Lyon Cedex

Chapter 5
Final Considerations and Open Scenarios

Abstract European river contract experiences demonstrate a growing integration between these contractual agreements and the other instruments of water resources management, and urban and territorial planning. Therefore, river contracts represent innovative *places* for a new governance of river ecosystems and territories, also in compliance with subsidiarity principle. The twofold nature of these contractual agreements—technical dimension and concertative approach—together with their expected wide evolution, allow to identify river contracts not only as sectoral tools for water resource protection and management, but also as catalysts of a new *culture of water*, recalling the deep interrelationships existing between hydrography, hydrogeology, ecology, sociology, economics, public health and cultural values.

The diffusion of river contracts (RC) in the European scenario is a phenomenon of great interest for the implementation of integrated water management policies. Starting from the first experiences in the 1980s, RC have acquired considerable flexibility and offered original solutions for problematic issues related to river basin management. The analysis of the experiences both completed and still underway illustrates a growing trend towards integration between RC and other instruments of basin management and urban and territorial planning.

Interest in RC heightened after the Second World Water Forum held in The Hague in 2000, where, for the first time, such contractual agreements were identified on a global level as suitable processes for promoting sustainable development of territories at the river basin scale. During the Forum, the formal definition of RC confirmed its relevance in terms of integrating the dimensions of public interest, economic performance, social values and environmental sustainability. The European Water Framework Directive (WFD), also in 2000, gave new impetus to water resources management, stressing the importance of appropriately organized forms of river basin management and participatory processes. Arguably, both the Second World Water Forum and the WFD have rendered the breeding ground fertile for the diffusion and adoption of RC as implementation tools for purposes of river basin planning, indeed through participatory and inclusive approaches.

M.L. Scaduto, *River Contracts and Integrated Water Management in Europe*, UNIPA Springer Series, DOI 10.1007/978-3-319-42628-0_5

Across the board, the underlying theme of the river basin as the reference unit for the implementation of integrated water management policies, has characterized all major cases of RC. This reference unit, however, often involves territories which are extremely complex and diverse from a geographic, environmental, social and political viewpoint, all features that can affect the breadth and scope of the RC. Notwithstanding the local environmental and geo-political territorial differences, the river basin has contributed to characterizing these voluntary agreements as highly concerted and inclusive instruments. In this sense, the all but secondary role that RC have taken on in restoring physiographic, administrative and management identities to river basins, has also contributed to helping rebuild up-stream and down-stream relationships, and orienting actions towards a territorial dimension modeled on the tenets of *bioregionalism* (Magnaghi 2011).

In this light, RC are capable of prompting new participatory processes that primarily constitute important forums for dialogue and knowledge sharing between public institutions, associations and local communities. Indeed, during the actual implementation of the RC, such venues for consultation are transformed into new forms of exercising governance over the territory and new ways of transposing European and national integrated water management policies into the different local contexts. For example, in the course of the dialogue among institutions and local stakeholders, RC may give rise to original innovations serving to generate more effective solutions for innovative water management policies (Allain 2010; Berry and Mollard 2010).

The interactions between public and private actors that may be achieved thanks to RC are virtually limitless, given the host of combinations possible among players and any number of forms of association between local authorities and individuals. In fact, State, regional and local institutions, as well as non-institutional stakeholders having some level of expertise, or that somehow interact in a given river ecosystem, may all partake in these contractual agreements. At the inter-municipal level, in particular, RC are able to promote extremely advantageous forms of association between the different local communities, especially where there is a need of consensus and cooperation, often due to geo-political fragmentation and/or low population density, abandonment phenomena of traditional productive activities and overall underdevelopment of the territory.

Thus, the subsidiarity principle too may find in the RC an important venue for its re-affirmation and concrete implementation, provided the institutionally agreed plan of actions be aptly designed with due consideration for local territorial identities and its realization take into account the different institutional competences and decision-making levels, revolving around a specific river basin/sub-basin.

RC are characterized in particular by a complex, essentially twofold nature whereby a technical and sectoral dimension coexists with the other of the concertative territorial governance (Bobbio and Saroglia 2008). These contractual and voluntary agreement are, therefore, capable of activating new processes of integration between river basin and spatial planning. This is one of the domains in which RC reveal their vast potential, in contrast both to the widespread structural weaknesses of the interrelationships between local systems of urban and territorial planning, and to the absence of true coordination between all sectoral instruments concerning rivers, and surface and underground waters management. In this sense, the RC is identified as a liaising tool between the two spheres of planning, as well as a solution to any possible overlap in institutional and legal competences of the actors involved, or between the actions and interventions envisaged by different programs and plans.

Undoubtedly, the specific potential inherent to RC must also come to terms with the different national and local regulatory frameworks that establish the degree of integration between integrated basin management and spatial planning. For example, in France the relations between *contrats de rivière* (*CdR*) and urban and territorial planning tools are decidedly strengthened by the constraint that the latter be coherent with the *Schémas Directeurs d'Aménagement et Gestion des Eaux* (*SDAGE*) and *Schémas d'Aménagement et Gestion des Eaux* (*SAGE*), as seen in the previous chapters. Therefore, basin plans give every assurance of effective synergy between the different sectoral programming instruments, representing the common regulatory framework of reference for both *CdR* and spatial planning tools.

The French example illustrates how RC prove instrumental in operatively and progressively assembling the functional mosaic of integrated basin planning and management, enabling an effective and lasting restructuring of an entire system of territorial, social, economic and cultural relations between up-stream and down-stream areas and communities. Moreover, in those contexts in which a *SAGE* is not yet in force, the reflection and actions underlying the implementation of a *CdR* may actually serve as the groundwork that sets the stage for the procedures that may bring to the adoption of a *SAGE*.

The analysis of experiences carried out in Europe, especially in France and Italy, clearly shows how RC should hardly be considered a mere sectoral tool, relevant only to the protection and management of water resources, but rather a generative process that spans the domains of hydrography, hydrogeology, ecology, sociology, economics, public health and culture.

Inasmuch as they constitute contractual agreements among public and private promoters and other participating stakeholders, RC can be tailored, depending on the case, to multiple fields of action relative to the thematic areas of specific local interest. For example, in addition to actions aimed at safeguarding and re-qualifying fluvial environments, initiatives for improving the quality standards of water resources may also be prompted along with others to sustain the production capacity of the agricultural, fishing and energy sectors. Similarly, purely technical measures related to contrasting geological risks can be coupled to initiatives seeking to bridge gaps in the overall knowledge base regarding specific hydrographic and territorial regions. For example, through an interdisciplinary approach, the structural and infrastructural dimension of a RC plan of actions may be integrated by conducting a census and determining the collective recognition of the component parts of individual entities of naturalistic interest (biodiversity, ecological networks, parks, reserves), anthropological and cultural interest (cultural heritage, historic built-up areas, cultural landscapes) and social interest (identity elements, gathering places, recreational areas), and their interdependencies (Magnaghi 2011). In particular, with reference to the cultural interrelations, which in fact underlie all others, it is interesting to point out how RC are able to build new awareness, on the part of local communities, by fostering a new *culture of water*, via a commitment to promoting education concerning ecosystems and forms of concerted dialogue, and, thus, to achieving a heightened collective awareness of the values underlying the water resources they share.

In this perspective, RC represent viable means for agencies and local communities to avail themselves of financial resources, otherwise seldom accessible, to be allotted for safeguarding and developing river territories and their natural and anthropic ecosystems as well as for promoting sustainable development, via apposite integrated action plans. In a number of cases its significant potential could bring to fruition integrated interventions for the comprehensive requalification of territorial areas, whether intensely populated or suburban, whereas the latter are otherwise all too commonly subject to processes of abandonment and marginalization, if anything.

Evidently, this inherent potential of RC must nevertheless come to terms with some principal limits that the analysis herein conducted on actual experiences in Europe has highlighted.

The reference to the territorial, ecologic and hydrogeological unit of the river basin, although addressing the natural context to which RC should refer, at times finds itself at odds with the concerns of an administrative, institutional, economic, social and political nature that distinctly characterize each territory. For example, hydrographic units of greater territorial extension may, per se, represent a limit to reaching adequate levels of internal cooperation amongst local actors and, thus, to achieving the planned objectives.

The complexity and the time scales of river dynamics are additional potentially critical points with respect to the application of RC, especially in relation to the total duration of the action plan. For example, the experiences in France prove that the average period of 5–7 years required for the implementation, of interventions may in some cases result inadequate for achieving the planned objectives above all in terms of contrasting geological risk and the environmental requalification of river contexts.

The most characterizing feature of RC is however still legal in nature and consists in the signing of the negotiated agreement on a voluntary basis. In other words, the territorial actors in a given hydrographic context are without any obligation to adhere to RC, although promoted in most cases by State and local institutions, thus only take part according to their specific interests and effective political biases. In this sense, RC as yet are devoid of any regulatory bearing unlike spatial planning and other integrated basin management tools. The non-binding nature of RC may result in partial success in terms of stakeholder involvement in a given river area, with the consequence of action plans being less than fully shared, if contrasted with cases enjoying full participation of private parties within the financial and planning framework of a RC.

The process of territorial consultation and the active participation of stakeholders are the two crucial factors in the implementation of RC. Almost without exception, the problems specific to a given local context require lengthy phases of shared analysis and demanding negotiations before reaching a definition of contents and objectives of the project, and the signing of the contract.

The time requirements for activating the contractual procedure, which almost invariably amount to approximately ten years, increase the overall probability of significant variations within the scenario of application of a given RC. Changes having most bearing can occur in terms of the composition of the partnership, the political representatives of the local authorities involved, the subject of interventions, the availability of financial resources and the natural dynamics of the hydrographic context. These consultation processes can prove drawn out and burdensome in terms of coordination of actors, conflict resolution and effective achievement of objectives.

Given the different evolution that RC have experienced in the European scenario, as well as the particularities of the experiences within each national context, the different renditions of such contractual agreements are directly linked to the legislative, institutional and local administration framework in matters of protection and management of water resources and of the territory. The differences in terms of normative references, procedures and contents, as well as the role of promoters and implementing subjects, may represent considerable limitations for the realization of RC impacting cross-border or interregional areas. For example, in the case of the *Segre River Contract*, whose hydrographic basin in extended across the border between France and Spain, in the central part of the Pyrenees, there were shortcomings in the effective coordination and true interaction among actors on either side of the border (Maury and Richard 2011). In actuality, for the realization of the *Segre River Contract*, signed in 2001, a managing body and a *comité de rivière* were instituted for each partner State, while each *comité de rivière* housed a college of representatives on behalf of its institutional counterpart. Although the synergy between the two managing entities and the corresponding *comité de rivière* proved instrumental in securing both local and European funding, the cross-border dimension was ultimately perceived more as a hindrance and a drawback than as an actual advantage. In this regard, the case of the *Segre River Contract* reveals the potential issues that may occur in the ambit of cross-border RC, in the presence of situations of contention linked to the contrasting uses of water resources, absent an effective degree of underlying institutional and cultural synergy among up-stream and down-stream stakeholders.

The example mentioned also confirms that features of political, administrative and technical steadfastness on the part of the organization that assumes the project manager role in a RC, represent other key aspects of the potential success of these negotiated and participatory processes. The crucial points, in this case, are closely tied to the actual capacity of the managing bodies to (I) implement, promote and support the dialogue between the parties involved, (II) coordinate the implementation of program of interventions, (III) be skilled in steering action plans throughout, and (IV) monitor, even *ex post*, the actual outcomes of interventions in the territories concerned. These aspects are particularly relevant in cases where the managing bodies are individual or groupings of local authorities, which may easily be subject to periodic variations of their political representatives and interlocutors, especially as a consequence of local rounds of voting. Conversely, State-instituted technical organisms, or those, in any case, lacking a predominant political component, are probably the best institutional candidates for the role of project manager in RC implementation. In Italy, for example, the State-instituted non-political bodies of the Basin Authorities, accountable for hydrographic units management, could take on a more decisive role in the promotion and implementation of RC, albeit in collaboration with the regional and local authorities, and in strict adherence to the principle of subsidiarity. This would tend to maximize the synergy between institutional competences, managerial, technical and financial capabilities, thus ensuring greater long-term administrative and management stability to activated RC.

As previously highlighted, RC can contribute to apply the principle of subsidiarity as they are operational tools capable of being tailored to the requirements of each hydrographic territory. In this sense, in order to stave off the emergence of conflicting situations, it is befitting that the politico-institutional dimension of a RC uphold this principle, thereby ensuring that planning and operational orientations actually be defined at appropriate administrative levels. In other words, the purported action plans must take into account the effective authority and decision-making powers of the lower-level government bodies amongst local institutions.

The principle of subsidiarity also refers to the theme of integration between integrated basin management and urban and territorial planning, although the interactions between the respective implementation tools remain beset with systemic frailty. In some contexts of application, however, the fact that the requirement of mutual compatibility between management tools still awaits a clear legislative and normative characterization, may entail difficulties in formalizing effective and synergistic relationships among RC, basin management plans and local spatial planning instruments. These substantial interdependences have to become something more than the mere, mutual mentions of the various tools and plans in their respective technical and normative documents.

The extent of actual integration among RC, basin plans and spatial planning tools reflects the internal consistency and, thus, validity of action program of each contractual agreement. In fact, whenever effective synergy between the various planning tools and a RC is maintained, the thematic structuring of the plan of interventions tends to be well-balanced and better integrated with the theoretical and operational orientations of basin plans, on the one hand, and with the plans for governing the territory at the regional, provincial or municipal levels, on the other. Furthermore, this specific integration aspect is all the more relevant in those contexts concerned with the issues of *Integrated Coastal Management* (Granit et al. 2014), recently promoted in Europe by Directive 2014/89/EC entitled *Directive Establishing a framework for maritime*, and *Community-based Coastal Management* (Harvey et al. 2001).

Another potentially crucial point in the current scenario of RC concerns the sources of funding and its appropriate use. At present, in most experiences implemented or still underway, the total investment required to bring the planned interventions to fruition has been guaranteed by public funding, while the financial contributions from the private sector have been quite limited. After all, given the inherent prevailing public interest in RC, it is quite understandable that the bulk of the budget derive from State and/or European funds allotted for Regional development and environmental expenditures. Under these circumstances, however, it is possible to focus on two potential pitfalls or areas of concern, the first regarding the redistribution of the allocated funds amongst the various implementing bodies, the other represented by the inherent potential for political speculation, in the form of tapping public resources for actions and interventions that are at times inconsistent, in part or in whole, with the objectives of a RC. In the first case, the greatest risks may present under untoward conditions of conflict arising between the actors involved or whenever there is any existing disproportion in the financial resources inequitably allocated among the various territorial contexts and social groups concerned. In the second case, the initiation of a RC could underlie local political interests, even in stark contrast with the purported goals of the RC paradigm and, if anything, more oriented to securing funds for interventions, all but entirely consistent with the ends of integrated management of river basins and matters pertaining thereto.

In light of the considerations expounded thus far, the realization of the potential of RC and, at the same time, the progressive troubleshooting of the crucial points, may actually be accomplished through (I) a greater legislative and financial support for their dissemination and implementation, (II) a deeper analysis of every undertaken RC experience and, consequently, a greater dissemination and sharing of consolidated knowledge and know-how, (III) true incentives for transferring methodological, technological and management skills matured in an increasingly unified European context, to the benefit of communities and territories within and without Europe.

Undoubtedly, at the European level the definition of an EC legislative and financial framework specifically dedicated to RC, appears as the first step towards a true evolution of these instruments. This innovation should clearly operate both in the sense of a greater normative alignment with other sectoral planning tools, as well as on a legal and regulatory plane by fine-tuning their contractual and participatory nature.

At present, the molding of such a European framework could already draw on the ample repertoire of RC, either completed or underway, as well as on relevant elements taken from various national and regional laws and regulation. In this respect, the issue of a specific EC directive seems a reasonable aspiration, specifically in terms of the regulation of RC and on the identification of appropriate funding sources. The transposition of such a directive would hasten the legal standardization of RC in all national contexts and the acknowledgment of their role in the implementation of integrated water management policies at the scale of the hydrographic basin.

The purpose of such a directive should also be to identify adequate programs and EC funding sources to support the dissemination and implementation of RC, especially to those contexts where they are as yet limited. On the financial front, the same directive could induce Member States to, in turn, identify the national and local budget sections to be allocated specifically for RC, also through eventual co-founding partnerships between public and private entities. Moreover, the establishment of European and national funds specifically earmarked for the implementation of water policies through RC could substantially contribute to limiting potential cases of political speculation aimed at securing public funds for interventions that are anything but wholly consistent with the ends of RC.

Another key issue for the evolution of the RC model, at the European level, consists in expanding the forms and vehicles for sharing knowledge, either consolidated or in the making, throughout the various national contexts. Specifically, it is a matter of fostering more effective analyses of the RC experiences both completed and still underway, through opportune methods, whether single or multiple, of systematization of knowledge frameworks, executive projects, shared and participatory processes, inclusive of the actual results achieved by each action program. Organizing such knowledge bases within national and regional information networks, while allowing access to them via a dedicated portal at the European level, is of utmost importance when one considers the prospect of more effective exchanges of best practices and know-how, both among different Member States and among the various regional and local contexts more closely concerned with the specifics of river basin management. The main goal, far from trying to uniform the variety and diversity of local declensions of RC according to a single abstract standard, should be, above all, to disseminate knowledge and permit mutual comparisons between the various local frameworks of reference. In all likelihood, only by providing potential promoters of RC with straightforward and effective access to the complete profiles of the experiences already implemented and those still underway, it will be possible to optimize the local declensions, while increasing the acceptance and adoption of such contractual instruments.

The considerations above make it seem only appropriate that an institutional and research network be activated, ideally comprising a European observatory together with similar observatories at each national level, for purposes of conducting advanced studies and promoting the diffusion of RC.

There are already some examples of active institutional observatories concerned with RC, such as the one operating at the national level in France, namely *Gest'eau* (http://www.gesteau.eaufrance.fr), and another at the regional level in Italy, to wit *Contrattidifiume.it* (http://www.contrattidifiume.it), established by the *Regione Lombardia*. Of particular interest are also two initiatives being conducted, as of late, in the Italian scenario, specifically the *National Board on River Contracts and* the *European Action Group* named *Participatory European Network on Water Governance—Smart Rivers Network*, promoted as of 2015 by the same National Board.

The *National Table on River Contracts* was established in 2007 as a national organism linked to the *Italian Coordination of Local Agendas 21*, with the aim of creating a venue for the exchange of best practices in integrated water management and for the promotion of RC in Italy. This body originally included the *Regione*

Umbria, several municipalities adhering to *Agenda 21*, local authorities which had already taken part in RC, environmental authorities responsible for managing fluvial waters as well as other associations. Before long, the *National Board* had received extensive approval from the State central administrations responsible for the management of water and the environment, regional and provincial administrations, and several municipal authorities, as well as from research centers and universities, professionals and trade associations. A first significant result was achieved in 2010, at the *5th Annual Meeting of the National Board* where the *National Charter of River Contracts* was presented as the declaration and official guidelines for the realization of new interventions aimed to re-qualify and enhance the quality status of river basins, through the application of the RC operational model.

At present, the *National Board on River Contracts* represents a valid prototype of a local observatory, characterized by the interaction of the skills and expertise of researchers, scholars, experts, technicians, professionals and representatives from the institutions, the research and academia communities, professional associations, as well as educational, civic and environmentalist groups. Under its roof, in fact, the modus operandi for promotion and dissemination incorporates the fundamental constituents of the methodological and procedural model of the RC, such as an interdisciplinary approach, active dialogue between the various stakeholders and regard for consultation and participatory processes.

From the experience of this *National Board* directly descended the second initiative, even more focused on the European scenario of RC. The above mentioned *European Action Group Participatory European Network on Water Governance—Smart Rivers Network*, in addition to members of the *Italian Board*, includes representatives and experts from other Member States. The intent of this working group is to promote forms of participatory governance throughout European hydrographic territories by creating a network for institutional cooperation, so as to promote greater awareness of the role of RC in the context of participatory processes applied to integrated water management policies, while expanding their dissemination to areas of more limited application, such as Eastern Europe. Currently, this emerging workgroup is part of the *Water Action Groups* within the framework of the *European Innovation Partnerships*. In this facet, it may hopefully pave the way for a new network of observatories on RC in Europe.

From the perspective of the next implementation of such a network of specialized observatories, it is also possible to identify, even at the technological level, some reference paradigms useful for more advanced sharing of consolidated knowledge, required for in-depth analyses of all cases in which RC have been applied. In fact, the realization of one or more technological platforms based on the potential of the web, the *GIS* (*Geographical Information System*) and related *webGIS* applications, the interoperability between computing systems and social-networking services for users, typical of social media, together may represent the optimal approach to achieve the objectives of sharing, disseminating and updating new common knowledge bases on RC.

Nowadays, the methodological and technological scenario offers applications of great interest and potential, for instance *Public Participation GIS* (*PPGIS*) that represent a particular adaptation of GIS-based technologies. Specifically, a PPGIS combines the management and mapping of geospatial and thematic information, typical of GIS applications, with full support of participatory processes relied on in many sectors, such as sectoral planning, project design and management of specific territorial facets of public interest (Brown and Kyttä 2014).

The PPGIS paradigm offers diverse features of interest for all those procedures which require, in the first instance, the mapping out and consolidation of general and thematic knowledge frameworks to be shared and collaboratively updated by numerous stakeholders having institutional authority and operational expertise, scientific and technical skills, and their own resources, differentiated according to their territorial roles as well as their political, social and economic interests. Indeed, the main potential of *PPGIS* lies in its ability to reach and involve a host of territorial actors, interest groups and even single individuals, in a manner adaptable to the single citizen profiles and skills of each, at the same time amplified due to the interactivity offered by such web 2.0-based tools.

As such, it is evident that PPGIS can be a valid methodological, technological and procedural solution to support the activation and promotion processes of RC, firstly for mapping out the knowledge frameworks and defining the plans of actions and, subsequently, for the actual implementation phases and the continuous monitoring of interventions. Therefore, having recourse to PPGIS within the ambit of the implementation process for RC could represent an apt alternative, especially for the engagement of local communities, and for the development of new forms of communication and sharing of knowledge and common languages, made accessible to different types of stakeholders in an easier and well-delineated way. In this perspective, PPGIS could productively collect through an integrated representation different procedures and practices that have already been developed during the implementation of numerous RC, and those that are still being developed within the ambit of national technical boards and observatories dedicated to this specific matter.

The possible forms of integration of PPGIS within the preparatory and implementation pathways of a given RC should include, first of all, activities of a methodological and technological nature shared among the promoters of the initiative and the other relevant players responsible for integrated basin management, spatial planning, Information Communication Technologies (ICT) and GIS, aimed at the implementation of dedicated public information platforms. Within these virtual collaborative spaces, general and thematic knowledge bases would then be opportunely systematized and updated in order to ensure that all players are allowed full access to the information base and to the different planning scenarios of the project, via dedicated web-based applications, also integrated with typical social media services.

As well-known advantages accruing from GIS-based technology are numerous, particularly regarding the logical organization and integration of data made available through various planning tools in use within a specific territory

(Zullo et al. 2015). Through GIS applications, horizontal and vertical relationships can be made explicit to all users, for example between basin management plans and spatial planning tools at its various levels of detail. Moreover, the opportunity offered by applying PPGIS for purposes of implementing RC could represent a welcome innovation with regard to the integration between water resources management and sectoral planning at the river basin scale, as well. Once an active and collaborative platform for the implementation of a RC were operational and running, in fact, users could participate, in agreement with the relevant institutions, in signaling areas of conflict arising between the different levels and tools of the sectoral planning, so as to advance hypotheses and eventual proposals for their solution to be incorporated into relevant programming and planning instruments.

After having consolidated the general and thematic knowledge bases within the *River Contract PPGIS*, the promoters would proceed to the first definition and complete digital representation of the plan of actions, ensuring it were shared and disseminated widely among all institutions and territorial communities, via the same PPGIS. This innovative modality of sharing and conveying the knowledge base would be followed by the second operational phase of the PPGIS platform implementation, focused on conducting a census so as to collate the various needs and observations on the part of both public and private territorial players. By way of accessible and user-friendly web-mapping tools, any private or legal entity could actually contribute to better defining the final project by integrating the initial knowledge base and the first draft of the RC plan, with their own critical considerations, comments and any other relevant information. This collaborative and inclusive *modus operandi* would be employed during all subsequent definition and approval stages of the final project and, later, throughout the implementation and constant monitoring of its program of interventions.

Arguably, having recourse to PPGIS, in and of itself, is hardly a guarantee of success for the realization of a RC, as is true for any other public participatory process, and should therefore be well pondered in relation to the overall project and to the territorial, institutional, socio-economic and cultural aspects of the concerned hydrographic territory. Especially in the case of a RC, during the development of the PPGIS platform, choices regarding the following variables are deemed crucial: (I) entities and stances participants will be asked to map, (II) the mapping technology and user interfaces adopted, (III) the methods and activities for interpreting the collected crowd-sourced data and knowledge. The overall result will obviously have to take into account the actual capacity of promoters to acknowledge and delegate an official consultation role to all those actors taking part in the participatory process, in order to reassure and motivate the participants also regarding the aware use of PPGIS as a tool for voicing and defending their perspectives and needs.

The methodological and technological option of implementing a River Contract PPGIS, therefore, could allow us to optimally attune (I) the requirements of mapping out and updating general and thematic interdisciplinary collaborative knowledge bases, (II) guarantees of constant access to public administrative and territorial information for all stakeholders, and (III) programs for sustaining the participatory processes of integrated management of hydrographic basins.

From a broader perspective at the European and global level, the introduction and the adoption of the technological and methodological paradigm of River Contract PPGIS, could represent an important catalyst for innovation in transferring sectoral skills, honed entirely via their completed or ongoing application, into different national contexts. In fact, the transferability issue, regards the management and procedural model of RC, takes on a pivotal role both in current and future scenarios, particularly with reference to cases concerning cross-border water basins, as well as in those wherein RC are part of international cooperation programs.

As regards the former of the two situations, the abovementioned example of the cross-border River Contract of the Segre (France-Spain) demonstrates how the degree of transferability can be crucial to the overall success of a RC. In fact, where the aim is to apply a particular declension of the instrument developed in a given local context, to different territorial, socio-economic and cultural contexts (even if sharing the identical hydrographic unit), the configuration of the new RC must take into account the methodological, managerial, procedural and technological elements effectively applicable to those very contexts. In the case of cross-border RC, moreover, it is believed that the transferability between the different local contexts involved assumes a major role, being able to facilitate a subsequent *replication effect* of the model in new local experiences regarding each individual national contexts.

For the more challenging cases of RC initiated and implemented within the framework of international cooperation programs, the transferability factor assumes a dimension of even greater complexity, because of the geographical, geo-political, legislative, regulatory, socio-economic and cultural diversities, as well as those pertaining to the financial resources respectively allocated by each cooperating partie. An interesting example, among the many experiences of this kind already implemented, is represented by the *Contrat de rivière de la Vallée du Sourou*, launched in 2003 for the homonymous hydrographic basin, located in West Africa. The project stems from an initiative promoted by the Belgian Walloon Region in cooperation with the State of Burkina Faso, subsequently included in the *Twin-Basin Project* sponsored by the *Réseau International des Organismes de Bassins* and by the *Office International de l'Eau* of Wallonia. The project, which took into account the guidelines for global cooperation defined in the *Fourth World Water Forum*, entailed transferring and adapting to the local reality the Walloon River Contract model experimented for the *Semois* river, essentially twinning the two respective territories and hydrographic basins (Rosillon 2007). The objectives that laid the foundations of the initiative included, firstly, reducing environmental degradation of the *Sourou* river basin, mostly caused by the extensive hydraulic transformations brought upon the region for purposes of increasing agricultural productivity, and, secondly, improving the conditions of life for the communities linked to the river itself. In this perspective, the RC has also become the application tool of the *Programme National de Lutte contre la desertification* launched by the State of Burkina Faso (Rosillon et al. 2005).

In the specific example recalled, the transfer of the RC paradigm resulted, initially, in the replication of the procedural and operational elements according to the *Contrat de Rivière du Semois* (i.e. implementation of the participatory process and consultation with local players; mapping out the knowledge base; instituting the *comités de rivière*; definition and implementation of the plan of actions). Obviously, the transfer of the RC model applied to the *Semois* basin required a suitable adaptation to the environmental conditions and to the issues relevant to local farming systems, consequently, the action plan was structured with specific attention to raising awareness regards environmental protection, to cope with ecological degradation within the *Sourou* Valley, to reducing agricultural crops within the *freedom spaces* of the river and to restoring riparian vegetation (Rosillon 2007).

On the whole, therefore, with reference to integrated water management, cross-border experiences and those regarding international cooperation clearly indicate the necessity for careful analysis of the elements of true transferability of RC between diverse geographical, geo-political, socio-economic and cultural contexts, as well as the importance of effective adaptability of the models being transferred to often disparate local contexts. Nevertheless, the versatility of RC for negotiation and consultation participatory processes clearly emerges, as well as the possibility of their dissemination to various contexts worldwide.

Of course, many factors and circumstances may contribute to some extent to advancing the dissemination of RC, still underway in Europe and throughout the world.

While the European and the global scenario of integrated water management policies keeps evolving, coming into closer alignment at the legislative, regulatory, legal, procedural and financial levels, the efforts of many actors, involved in various capacities and functions in the protection, enhancement and management of water resources, are pursuing a deeper understanding of the potential, current limitations and especially foreseeable future developments of RC.

With these aims, the research illustrated herein intends to stimulate scientific and institutional debate and share considerations on the breadth of the issues regarding RC actual applications, the salient aspects of which have been outlined in this and the previous chapters.

The arguments are thus submitted for further confrontation with the ensuing developments of the institutional, scientific and technological debates as regards the actual impact of RC on river basins and their territories safeguard and management.

At present, the greatest expectations for the near future remain those of a general acknowledgement of RC, by numerous countries and institutions, as implementation tools of water management policies and, in particular, of basin management policies. At the same time, the research and experiments being conducted in Europe and worldwide are focusing on the likely contribution that will accrue from such instruments of negotiation, consultation and participation in terms of ensuring the necessary integration process between -spatial planning and integrated management of water resources at the hydrographic basin scale.

The hope of the author is that the considerations and proposals advanced in this research work might inspire updates and new insights concerning the matter, perhaps contributing to stimulate debate on the multi-faceted relationships between river contracts, protection of fluvial ecosystems, sustainable management of water resources and innovative development of hydrographic territories.

References

Allain S (2010) Social participation in french water management: contributions to River Basin Governance and new challenges. In: Berry KA, Mollard E (eds) Social participation in Water Governance and management critical and global perspectives. Sterling, VA, London, pp 95–114

Berry KA, Mollard E (2010) Social participation in Water Governance and management critical and global perspectives. Sterling, VA, London

Bobbio L, Saroglia P (2008) Lungo il fiume delle politiche contrattualizzate. Dall'esperienza francese al caso piemontese. Laboratorio di Politiche-Corep, Torino, pp 1–28

Brown G, Kyttä M (2014) Key issues and research priorities for public participation GIS (PPGIS): A synthesis based on empirical research. Appl Geogr 46:122–136

Granit J, Lymer B L, Olsen S, Lundqvist J, Lindström A (2014) Water governance and management challenges in the continuum from land to the coastal sea—spatial planning as a management tool. Paper 22, Stockholm International Water Institute (SIWI)

Harvey N, Clarkea BD, Carvalho P (2001) The role of the Australian Coastcare program in community-based coastal management: a case study from South Australia. Ocean Coast Manage 44:161–181

Magnaghi A (2011) Contratti di fiume e pianificazione: uno strumento innovativo per il governo del territorio. In: Bastiani M (ed) Contratti di Fiume. Pianificazione strategica e partecipata dei bacini idrografici. Approcci, esperienze, casi studio. Dario Flaccovio Editore, Palermo, pp 31–44

Maury C, Richard S (2011) La difficile gestion de l'eau en contexte transfrontalier: un exemple franco-espagnol. Articulo. J Urban Res (Online) 6

Rosillon F (2007) Contribution à la gestion intégrée des eaux et des sols à travers l'application du contrat de rivière Sourou au Burkina Faso. Actes des JSIRAUF, Hanoi, 6–9 novembre 2007

Rosillon F, Vander Borght P, Bado-Sama H (2005) River contract in Wallonia (Belgium) and its application for water management in the Sourou valley (Burkina Faso). Water Sci Technol 52(9):85–93

Zullo F, Ciabò S, Fiorini L, Marucci A, Olivieri S, Perazzitti S, Romano B (2015) Multilevel planning regional management. A GIS Platform Structure. In: Gambardella C (ed) Heritage and Technology Mind Knowledge Experience Le Vie dei Mercanti _ XIII Forum Internazionale Di Studi. La Scuola di Pitagora srl, Napoli, pp 363–371

FSC
www.fsc.org
MIX
Papier | Fördert
gute Waldnutzung
FSC® C083411

Kolibri 360 GmbH
Ferdinand-Jühlke-Straße 7
99095 Erfurt, Deutschland
produktsicherheit@kolibri360.de